LONDON MATHEMATICAL SOCIETY LECTURE N

Editor: PROFESSOR G. C. SHEPHARD, Univers

This series publishes the records of lectures and s... topics in mathematics held at universities throughout the world. For the most part, these are at postgraduate level either presenting new material or describing older material in a new way. Exceptionally, topics at the undergraduate level may be published if the treatment is sufficiently original.

Prospective authors should contact the editor in the first instance.

Already published in this series

1. General cohomology theory and K-theory, PETER HILTON.
2. Numerical ranges of operators on normed spaces and of elements of normed algebras, F. F. BONSALL and J. DUNCAN.
3. Convex polytopes and the upper bound conjecture, P. McMULLEN and G. C. SHEPHARD.
4. Algebraic topology: A student's guide, J. F. ADAMS.
5. Commutative algebra, J. T. KNIGHT.
6. Finite groups of automorphisms, NORMAN BIGGS.
7. Introduction to combinatory logic, J. R. HINDLEY, B. LERCHER and J. P. SELDIN.
8. Integration and harmonic analysis on compact groups, R. E. EDWARDS.
9. Elliptic functions and elliptic curves, PATRICK DU VAL.
10. Numerical ranges II, F. F. BONSALL and J. DUNCAN.
11. New developments in topology, G. SEGAL (ed.).
12. Proceedings of the Symposium in complex analysis Canterbury, 1973, J. CLUNIE and W. K. HAYMAN (eds.).
13. Combinatorics, Proceedings of the British combinatorial conference, 1973, T. P. McDONOUGH and V. C. MAVRON (eds.).
14. Analytic theory of abelian varieties, H. P. F. SWINNERTON-DYER.
15. Introduction to topological groups, P. J. HIGGINS.
16. Topics in finite groups, TERENCE M. GAGEN.
17. Differentiable germs and catastrophes, THEODOR BRÖCKER and L. C. LANDER.

London Mathematical Society Lecture Note Series. 18

A Geometric Approach to Homology Theory

by S. BUONCRISTIANO,
C.P. ROURKE,
and B.J. SANDERSON

CAMBRIDGE UNIVERSITY PRESS
CAMBRIDGE
LONDON · NEW YORK · MELBOURNE

Published by the Syndics of the Cambridge University Press
The Pitt Building, Trumpington Street, Cambridge CB2 1RP
Bentley House, 200 Euston Road, London NW1 2DB
32 East 57th Street, New York, N. Y. 10022, USA
296 Beaconsfield Parade, Middle Park, Melbourne 3206, Australia

Library of Congress Catalogue Card Number: 75-22980

ISBN: 0 521 20940 4

Printed in Great Britain
at the University Printing House, Cambridge
(Euan Phillips, University Printer)

BST
R

Contents

Introduction

The purpose of these notes is to give a geometrical treatment of generalised homology and cohomology theories. The central idea is that of a 'mock bundle', which is the geometric cocycle of a general cobordism theory, and the main new result is that any homology theory is a generalised bordism theory. Thus every theory has both cycles and cocycles; the cycles are manifolds, with a pattern of singularities depending on the theory, and the cocycles are mock bundles with the same 'manifolds' as fibres.

The geometric treatment, which we give in detail for the case of pl bordism and cobordism, has many good features. Mock bundles are easy to set up and to see as a cohomology theory. Duality theorems are transparent (the Poincaré duality map is the identity on representatives). Thom isomorphism and the cohomology transfer are obvious geometrically while cup product is just 'Whitney sum' on the bundle level and cap product is the induced bundle glued up. Transversality is built into the theory - the geometric interpretations of cup and cap products are extensions of those familiar in classical homology. Coefficients have a beautiful geometrical interpretation and the universal coefficient sequence is absorbed into the more general 'killing' exact sequence. Equivariant cohomology is easy to set up and operations are defined in a general setting. Finally there is the new concept of a generalised cohomology with a sheaf of coefficients (which unfortunately does not have all the nicest properties).

The material is organised as follows. In Chapter I the transition from functor on cell complexes to homotopy functor on polyhedra is axiomatised, the mock bundles of Chapter II being the principal example. In Chapter II, the simplest case of mock bundles, corresponding to pl cobordism, is treated, but the definitions and proofs all generalise to the more complicated setting of later chapters. In Chapter III is the geometric treatment of coefficients, where again only the simplest case,

pl bordism, is treated. A geometric proof of functoriality for coefficients is given in this case. Chapter IV extends the previous work to a generalised bordism theory and includes the 'killing' process and a discussion of functoriality for coefficients in general (similar results to Hilton's treatment being obtained). In Chapter V we extend to the equivariant case and discuss the $\mathbf{Z}_2$ operations on pl cobordism in detail, linking with work of tom Dieck and Quillen. Chapter VI discusses sheaves, which work nicely in the cases when coefficients are functorial (for 'good' theories or for 2-torsion free abelian groups) and finally in Chapter VII we prove that a general theory is geometric. The principal result is that a theory has cycles unique up to the equivalence generated by 'resolution of singularities'. The result is proved by extending transversality to the category of CW complexes, which can now be regarded as geometrical objects as well as homotopy objects. Any CW spectrum can then be seen as the Thom spectrum of a suitable bordism theory. The intrinsic geometry of CW complexes, which has strong connections with stratified sets and the later work of Thom, is touched on only lightly in these notes, and we intend to develop these ideas further in a paper.

Each chapter is self-contained and carries its own references and it is not necessary to read them in the given order. The main pattern of dependence is illustrated below.

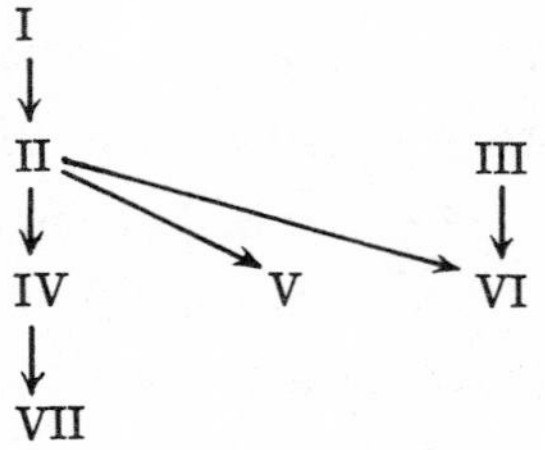

The germs of many of the ideas contained in the present notes come from ideas of Dennis Sullivan, who is himself a tireless campaigner for the geometric approach in homology theory, and we would like to dedicate this work to him.

NOTE ON INDEXING CONVENTIONS

Throughout this set of notes we will use the opposite of the usual convention for indexing cohomology groups. This fits with our geometric description of cocycles as mock bundles - the dimension of the class then being the same as the fibre dimension of the bundle. It also means that coboundaries reduce dimension (like boundaries), that both cup and cap products add dimensions and that, for a generalised theory, $h^n(\text{pt.}) \cong h_n(\text{pt.})$. However the convention has the disadvantage that ordinary cohomology appears only in negative dimensions. If the reader wishes to convert our convention to the usual one he has merely to change the sign of the index of all cohomology classes.

I·Homotopy functors

The main purpose of this chapter is to axiomatise the passage from functors defined on pl cell complexes to homotopy functors defined on polyhedra. Principal examples are simplicial homology and mock bundles (see Chapter II).

Our main result, 3.2, states that the homotopy category is isomorphic to the category of fractions of pl cell complexes defined by formally inverting expansions. Thus to define a homotopy functor, it is only necessary to check that its value on an expansion is an isomorphism. Analogous results for categories of simplicial complexes have been proved by Siebenmann, [3].

In §4, similar results are proved for Δ-sets. This gives an alternative approach to the homotopy theory of Δ-sets (compare [2]).

In §6 and §7, we axiomatise the construction of homotopy functors and cohomology theories. Here we are motivated by the coming application to mock bundles in Chapter II, where the point of studying cell complexes, rather than simplicial complexes, becomes plain as the Thom isomorphism and duality theorems fall out.

The idea of using categories of fractions comes (to us) from [1] where results, similar to those contained in §4 here, are proved.

Throughout the notes we use basic pl concepts; for definitions and elementary results see [4, 6 or 8].

1. DEFINITIONS

Ball complexes

Let K be a finite collection of pl balls in some $\mathbf{R}^n$, and write $|K| = \cup\{\sigma : \sigma \in K\}$. Then K is a <u>ball complex</u> if

(1) $|K|$ is the disjoint union of the interiors $\mathring{\sigma}$ of the $\sigma \in K$, and

(2) $\sigma \in K$ implies the boundary $\dot{\sigma}$ is a union of balls of K.

It then follows that

(3) if $\sigma, \tau \in K$, then $\sigma \cap \tau$ is a union of balls of K.

Notice that we do not assume $\sigma \cap \tau \in K$.

A subset $L \subset K$ is a subcomplex if L is itself a ball complex, and we write (K, L) for such a pair. If (K_0, L_0) is another pair, and $K_0 \subset K$, $L_0 \subset L$ are subcomplexes, then there is the inclusion $(K_0, L_0) \subset (K, L)$. An isomorphism $f : (K, L) \to (K_1, L_1)$ is a pl homeomorphism $f : |K| \to |K_1|$ such that $f|L| = |L_1|$, and $\sigma \in K$ implies $f(\sigma) \in K_1$. In the case where K and K_1 are simplicial complexes, there are simplicial maps $f : (K, L) \to (K_1, L_1)$. The product $K \times L$ of ball complexes K, L is defined by $K \times L = \{\sigma \times \tau \mid \sigma \in K, \tau \in L\}$.

The categories Bi and Bs

Now define the category Bi to have for objects pairs (K, L) and morphisms generated by isomorphisms and inclusions (i. e., a general morphism is an isomorphism onto a subpair). The category Bs has the same objects but the generating set for the morphisms is enlarged to include simplicial maps between pairs of simplicial complexes.

Subdivisions

If L', L are ball complexes with each ball of L' contained in some ball of L and $|L'| = |L|$, we say L' subdivides L, and write $L' \lhd L$. The categories Bi and Bs enjoy a technical advantage over categories of simplicial complexes; namely, if $L \subset K$ and $L' \lhd L$, then there is a complex $L' \cup K = \{\sigma : \sigma \in L'$ or $K - L\}$.

Collapsing

We assume familiarity with the notion of collapsing, as in [6], for example. Suppose $(K_0, L_0) \subset (K, L)$, where $L_0 = L \cap K_0$; then we have a collapse $(K, L) \searrow (K_0, L_0)$ if $K \searrow K_0$ and any elementary collapse in the sequence from a ball in L is across a ball in L (so that in particular, $L \searrow L_0$). We call the inclusion $(K_0, L_0) \subset (K, L)$ an expansion. The composition of an expansion with an isomorphism is still called an expansion.

2. SUBDIVISION IN THE CATEGORY OF FRACTIONS

Let $B = Bi$ or Bs, and let Σ denote the set of expansions. The category of fractions $B[\Sigma^{-1}]$ is formed by formally inverting expansions. Thus the objects are the same. New morphisms e^{-1}, $e \in \Sigma$, are introduced, and a morphism in the category of fractions is then an equivalence class of formal compositions $g_1 \circ g_2 \circ \dots \circ g_n$, where $g_i \in B$ or $g_i = e_i^{-1}$ for some $e_i \in \Sigma$. The equivalence relation is generated by the following operations:

(i) replace h by $f \circ g$ if $h = fg$ and $f, g \in B$;

(ii) introduce $e \circ e^{-1}$ or $e^{-1} \circ e$, $e \in \Sigma$;

(iii) replace $(e_2 e_1)^{-1}$ by $e_1^{-1} \circ e_2^{-1}$.

In fact operation (iii) is a consequence of operations (i) and (ii). Denote the equivalence class of a formal composition by $\{g_0 \circ g_1 \circ \dots \circ g_n\}$.

The category of fractions is characterised by a universal mapping property; namely, given any functor $F : B \to C$ such that $F(e)$ is an isomorphism for each $e \in \Sigma$, then there exists a unique functor F' so that

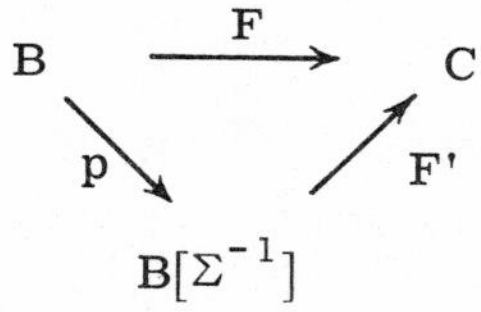

commutes, where p is the natural map.

For simplicity, in the rest of the paper we will ignore pairs (K, L) with $L \neq \emptyset$ when the general case can be obtained by making minor adjustments. We first observe that any morphism in $Bi[\Sigma^{-1}]$ may in fact be written $\{e^{-1} \circ f\}$ by repeated use of the following lemma.

Lemma 2.1. *Let* $e : J \to K$ *be an expansion and* $f : J \to L$ *a morphism in* Bi. *Then there is an expansion* e_0 *and morphism* f_0 *so that*

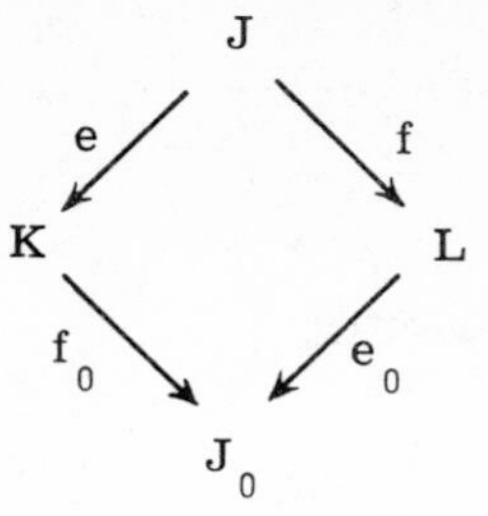

commutes.

Proof. Define $J_0 = K \cup_J L$, where J is regarded as a subcomplex of both by e and f. Then let e_0 and f_0 be the obvious inclusions. That e_0 is an expansion follows by echoing the collapse $K \searrow J$.

Remark. The lemma fails for Bs. For instance, take $K = I \times I$, $J = \{0\} \times I$, and $L = \{0\}$. Then f_0 must be degenerate on eJ and hence be degenerate on the 2-cell in K.

Now suppose $L' \lhd L$; then there are inclusions $i : L \to L \times I \cup L' \times \{1\}$ and $e : L' \to L \times I \cup L' \times \{1\}$. Then e is an expansion, and so we have a morphism $\{e^{-1} \circ i\} : L \to L'$ in $B[\Sigma^{-1}]$, called *subdivision* and denoted $\rhd(L, L')$.

Lemma 2.2. $\rhd$ *is functional; that is,*

(i) $\rhd(L, L) =$ *identity,*

(ii) *if* $L'' \lhd L' \lhd L$, *then*

$$\rhd(L', L'') \circ \rhd(L, L') = \rhd(L, L'').$$

Proof. For part (i), we must show that if $i_0, i_1 : K \to K \times I$ are the inclusions, then $\{i_0\} = \{i_1\}$. This is proved by simple collapsing arguments. First, consider $K \times I$ subdivided to $(K \times I)'$ by placing K at $K \times \{\frac{1}{2}\}$. There are then inclusions $i'_0, i'_1, i_{\frac{1}{2}}$ of K in $(K \times I)'$ and a reflection $r : (K \times I)' \to (K \times I)'$ about the half-way level. Now $ri_{\frac{1}{2}} = i_{\frac{1}{2}}$, and $\{i_{\frac{1}{2}}\}$ is an isomorphism since $(K \times I)' \searrow K \times \{\frac{1}{2}\}$. It follows that $\{r\} =$ identity. Since $ri'_0 = i'_1$, we have $\{i'_0\} = \{i'_1\}$. The result then follows by considering $K \times \Delta^2$. This argument is essentially due to Siebenmann [3; p. 480].

For part (ii), let Δ^2 be a 2-simplex with vertices v_0, v_1, v_2 and opposite faces $\Delta^1_0, \Delta^1_1, \Delta^1_2$. Let $(L \times \Delta^2)' = L \times \Delta^2 \cup L' \times \Delta^1_0 \cup L'' \times \{v_2\}$. Then $(L \times \Delta^2)' \searrow L' \times \Delta^1_0 \cup L \times \Delta^1_1 \cup L'' \times \{v_2\} \searrow L'' \times \{v_2\}$. The proof is completed by a diagram chase. See Fig. 1.

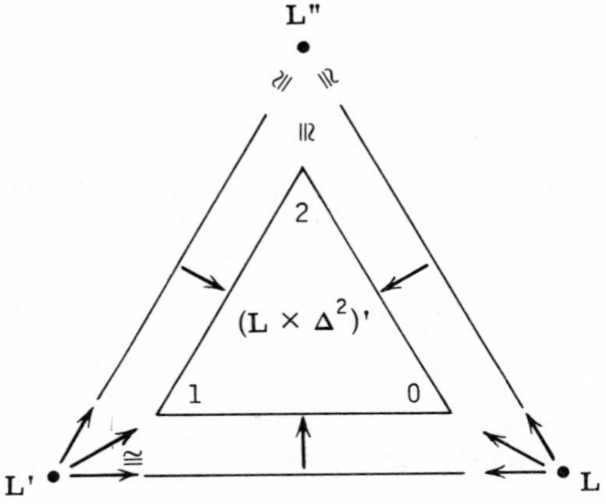

Fig. 1

Lemma 2.3. <u>Suppose given</u> $L' \lhd L$; <u>then there exists</u> $L'' \lhd L'$ <u>such that</u> $(L \times I) \cup L'' \times \{1\} \searrow L \times \{0\}$.

Proof. Let $\sigma_1, \ldots, \sigma_n$ be the balls of L listed in order of decreasing dimensions. We subdivide L in n steps. After step r, the interiors of $\sigma_1, \ldots, \sigma_r$ are not touched again. Suppose r-steps completed, and let $\sigma = \sigma_{r+1}$. Then σ has been subdivided to σ', say. Let $\tau \in \sigma'$ be a cell with $\dim \tau = \dim \sigma$, and $\tau \cap \dot{\sigma} = \emptyset$. If no such τ exists, perform a preliminary subdivision of σ. Now assume there is such a τ. Then $\sigma_c = \sigma$ - interior (τ) is a collar on $\dot{\sigma}$ in σ, and $\sigma_c \subset \sigma'$. Subdivide σ_c to σ''_c so that a collar projection $\sigma''_c \to \dot{\sigma}''$ is simplicial. $\sigma'' = \tau \cup \sigma''_c$ is the required subdivision of σ. Note the cylindrical collapse $\sigma''_c \searrow \dot{\sigma}''$.

The resulting complex L" clearly has the desired property since

$$(\sigma\times I)\cup\sigma''\times\{1\}\searrow(\dot{\sigma}\times I)\cup\sigma\times\{0\}\cup\sigma''_c\times\{1\}\searrow(\dot{\sigma}\times I)\cup\sigma\times\{0\},$$

where the first collapse is elementary, and the second is cylindrical.

Corollary 2.4. *Suppose* $L'\lhd L$; *then* $\rhd(L, L') : L\to L'$ *is an isomorphism.*

Proof. It follows from 2.3 that there is $L''\lhd L'$ and $L'''\lhd L''$, so that $\rhd(L, L'')$ and $\rhd(L', L''')$ are isomorphisms. The result now follows from 2.2.

3. ISOMORPHISM WITH THE HOMOTOPY CATEGORY

Now let Bh denote the category with objects pairs of ball complexes, and morphisms homotopy classes of continuous maps. Then there are natural maps

$$\begin{array}{ccc} \mathrm{Bi} & \searrow & \\ \cap & & \mathrm{Bh}\,, \\ \mathrm{Bs} & \nearrow & \end{array}$$

and by the universal mapping property, we have a diagram

$$\begin{array}{ccc} \mathrm{Bi}[\Sigma^{-1}] & \xrightarrow{\ \alpha\ } & \\ \beta\downarrow & & \mathrm{Bh}\,. \\ \mathrm{Bs}[\Sigma^{-1}] & \xrightarrow{\ \gamma\ } & \end{array} \tag{3.1}$$

Theorem 3.2. *The maps in diagram (3.1) are isomorphisms of categories.*

Proof. Since all three categories have the same objects, it suffices to show that each of α, β, γ is an isomorphism on the set of morphisms from K to L for any K and L. We prove this in three steps:

A. α is surjective,

B. α is injective,

C. β is surjective.

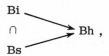

The result then follows by commutativity. We first observe that α and γ are compatible with subdivision; i. e.,

Remark 3. 3. Suppose $L' \lhd L$; then $\alpha(\rhd(L, L'))$ is the homotopy class of the identity map.

Step A. α is surjective.

Suppose $[f] : K \to L$ is a homotopy class; then by the pl approximation theorem, * there exist subdivisions $K' \lhd K$, $L' \lhd L$, and a simplicial map $f' : K' \to L'$ so that $[f'] = [f]$. Let $M(f')$ be the simplicial mapping cylinder of f'. There is then an inclusion $i : K' \to M(f')$, and an expansion $e : L' \to M(f')$. It follows from 3. 3 that

$$[f] = \alpha(\rhd (L, L')^{-1} \{e^{-1} \circ i\} \rhd (K, K')).$$

Step B. α is injective.

By 3. 1, it is sufficient to show that for any diagram

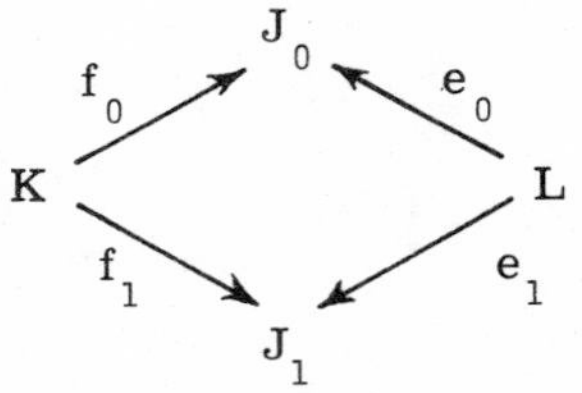

in which the e_i are expansions, and such that $\alpha\{e_0^{-1} \circ f_0\} = \alpha\{e_1^{-1} \circ f_1\}$, we have $\{e_0^{-1} \circ f_0\} = \{e_1^{-1} \circ f_1\}$. Now $\alpha\{e_0^{-1} \circ f_0\} = \alpha\{e_1^{-1} \circ f_1\}$ means that the diagram homotopy commutes, provided we regard e_0 and e_1 as homotopy equivalences.

Now let $J = J_0 \cup_L J_1$. Then by homotopy commutativity and the relative pl approximation theorem (see footnote), there is a simplicial map $f : (K \times I)' \to J'$ so that $f \,|\, |K| \times \{i\} = f_i$, $i = 0, 1$. Now consider the following diagram in which arrows marked e are expansions, and arrows marked $\rightarrowtriangle$ are compositions of subdivision followed by inclusion:

---

* This is weaker than the usual simplicial approximation theorem. See [4] or [8] for a short proof.

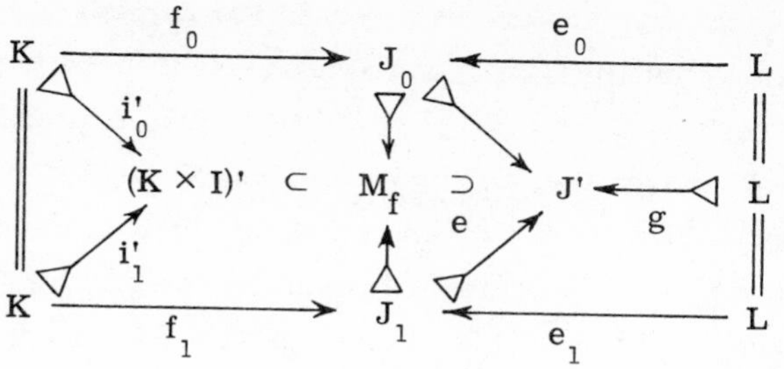

Commutativity in $Bi[\Sigma^{-1}]$ follows from definitions and 2.2. (Note that the left-hand half commutes by 2.2(i).) The map g is an isomorphism by 2.2 and 2.4.

The result follows.

<u>Step C.</u> β is surjective.

It is sufficient to show that the class $\{f\}$ of a simplicial map $f : K \to L$ is in the image, since any map in Bs is a composition of maps in Bi and simplicial maps. Consider the commutative diagram

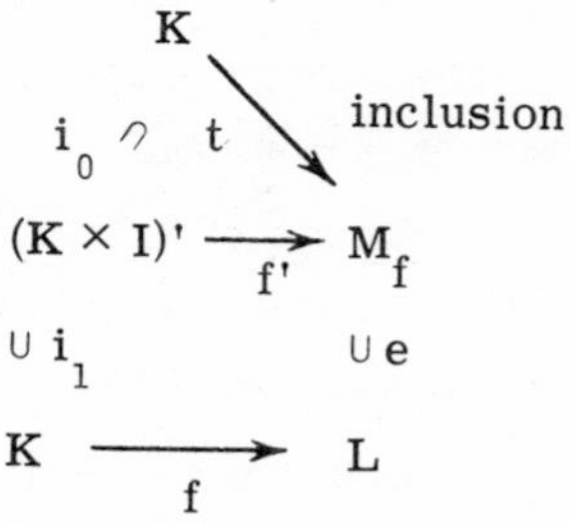

where M_f is the simplicial mapping cylinder of f, and $(K \times I)'$ is $K \times I$ derived at $K \times \{\frac{1}{2}\}$. The simplicial map f' is defined using the obvious vertex map. Then i_0 and e are expansions, and we have $\{f\} = \{e^{-1} \circ f' \circ i_1\} = \{e^{-1} \circ t \circ i_0^{-1} \circ i_1\} = \beta\{e^{-1} \circ t \circ i_0^{-1} \circ i_1\}$, since each map in the bracket lies in Bi.

4. Δ-SETS AND THE CATEGORY OF FRACTIONS

We assume familiarity with the basic definitions involving Δ-sets which are found in [2]. An inclusion of Δ-sets $L \subset K$ is called an <u>elementary expansion</u> if K is obtained from L by attaching a set of simplexes $\Delta^n(s)$, for s in some indexing set, to L via Δ-maps $f_s : \Lambda_{n,i}(s) \to L$. An inclusion $L \subset K$ is an <u>expansion</u> if there is a countable sequence $L \subset L_1 \subset L_2 \subset L_3 \ldots$ of elementary expansions so that $K = \cup L_i$. In particular, the 'end inclusions' of K in $K \otimes I$ are easily seen to be expansions. Let Σ denote the set* of Δ-maps which are expansions, and let $\underline{\Delta}[\Sigma^{-1}]$ denote the resulting category of fractions. The analogue of 2.1 is easily proved for Δ-sets so that any morphism in $\underline{\Delta}[\Sigma^{-1}]$ may be written $\{e^{-1} \circ f\}$. Let $\underline{\Delta}h$ denote the category of Δ-sets and homotopy classes of continuous maps between realisations.

Theorem 4.1. <u>The canonical functor $\underline{\Delta}(\Sigma^{-1}) \to \underline{\Delta}h$ is an isomorphism.</u>

Proof. The proof is analogous to the proof of 3.2. The main modifications are to replace the cell subdivisions used by simplicial ones. This can be done by deriving. Also, the Δ-sets need to be replaced by simplicial complexes before applying the simplicial approximation theorem.† This is done by using $(L \times I)'$ with $L \times \{1\}$ derived twice, and $L \times \{\frac{1}{2}\}$ derived once. We leave the reader to check details.

Now let K be a Kan Δ-set, and let $[X, K]$ denote Δ-homotopy classes of Δ-maps $X \to K$; let $\{X, K\}$ denote the set of morphisms $X \to K$ in $\underline{\Delta}(\Sigma^{-1})$. From Theorem 4.1, we have a well-defined function $\psi : [X, K] \to \{X, K\}$. In fact, ψ can easily be well-defined without the aid of 4.1, and the proof of the following theorem is then independent of 4.1.

Theorem 4.2. <u>Suppose X is a Δ-set and K is a Kan Δ-set. Then the canonical function</u>

$$\psi : [X, K] \to \{X, K\}$$

<u>is a bijection.</u>

---

* P. May has pointed out that there are set theoretic problems here. We adopt MacLane's axiom of one universe [7; p. 22]. Theorem 4.1 then shows that $\underline{\Delta}[\Sigma^{-1}]$ is a category in the universe.

† The required approximation theorem is given in [2; 5.1].

Proof. We first show that ψ is surjective. An element of $\{X, K\}$ has a representative $e^{-1} \circ f$, where f is a Δ-map and e is an expansion. Using the Kan condition on K, we find a Δ-map r so that $r \circ e = \mathrm{id}$. Define $g = r \circ f$. Then $\{g\} = \{r \circ f\} = \{r \circ e \circ e^{-1} \circ f\} = \{e^{-1} \circ f\}$.

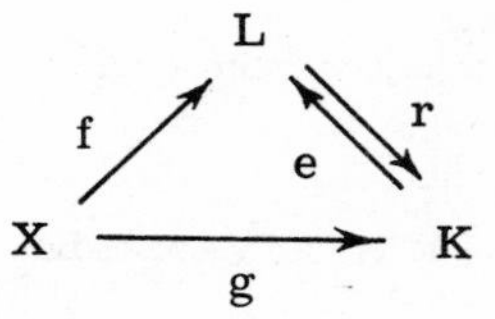

Thus $\psi[g] = \{e^{-1} \circ f\}$. To see that ψ is injective, suppose that $\psi[g_0] = \psi[g_1]$. This means that g_1 is obtained from g_0 by a sequence of the following steps and their inverses:

(i) replace (fg) by $f \circ g$,

(ii) introduce $e^{-1} \circ e$ or $e \circ e^{-1}$,

(iii) replace $(e_2 e_1)^{-1}$ by $e_1^{-1} \circ e_2^{-1}$.

Now it is easy to see that the map g defined above is unique up to Δ-homotopy; use the expansion $L \times \{0\} \cup K \otimes I \cup L \times \{1\} \to L \otimes I$. Hence after each step (ii), we can map each Δ-set in the composition into K uniquely up to Δ-homotopy. It follows that g_0 and g_1 are the same up to Δ-homotopy.

Now let CWh denote the category of CW complexes and homotopy classes of maps. Then there is the composition

$$f : \underline{\Delta}[\Sigma^{-1}] \to \underline{\Delta}h \to \mathrm{CWh}.$$

Theorem 4. 3. $f : \underline{\Delta}[\Sigma^{-1}] \to \mathrm{CWh}$ is an equivalence of categories.

Proof. It is sufficient to show that if X is a CW complex, then the natural map $|S(X)| \to X$ is a homotopy equivalence. From definitions, we have an isomorphism $\beta \circ \alpha$:

$$\pi_n SX \xrightarrow{\alpha} \pi_n |SX| \xrightarrow{\beta} \pi_n X.$$

But α is an isomorphism by a relative version of 4. 2. It follows that β is an isomorphism, and the result follows from J. H. C. Whitehead's theorem [5].

Remark 4. 4. There are easily provable relative versions of the theorems presented in this section.

5. HOMOTOPY FUNCTORS

We rephrase Theorem 3. 2 in terms of functors defined on $B = Bi$ or Bs. Let $T : B \to C$ be a functor.

Axiom C (Collapse). Let $e : (K_0, L_0) \to (K, L)$ be an expansion. Then $T(e)$ is an isomorphism.

From the universal property of the category of fractions and Theorem 3. 2, we have

Theorem 5. 1. Any functor T satisfying C factors uniquely through Bh; moreover, if T is defined on Bi, then it extends uniquely to Bs.

Suppose now T is contravariant and satisfies Axiom C. Denote the canonical extension by T also. Suppose $|L| = |K|$; then we have an isomorphism

$$T[id] : T(L) \to T(K).$$

If $L \lhd K$, we call this an amalgamation isomorphism and write

$$am : T(L) \to T(K).$$

From definitions, we see that $am = T(\rhd (K, L))$. The inverse isomorphism is called a subdivision isomorphism and we write

$$sd : T(K) \to T(L).$$

Now let Ph be the category of compact polyhedra and homotopy classes of continuous maps. Let $T : Bi \to S_*$ be a functor satisfying Axiom C, where S_* is the category of based sets. We define $T : Ph \to S_*$ as follows.

An element of $T(P)$ is an equivalence class of elements of $T(K)$, where $|K| = P$. Suppose $|K_i| = P$, $i = 0, 1$. Then $u_0 \in T(K_0)$ is defined to be equivalent to $u_1 \in T(K_1)$ if there exists $u \in T(K')$ for some $K' \lhd K_i$, such that $am(u_i) = u$, $i = 0, 1$. Alternatively, $T(P) = \varinjlim T(K)$ where the limit is taken over all K with $|K| = P$. The canonical function $T(K) \to T(P)$ is of course a bijection. If $[f] : P \to Q$ is a homotopy class, then $T[f]$ is defined by choosing a representative for $[f]$ in Bs.

Finally, note that for an arbitrary category C, we could define $T : Ph \to C$ by choosing for each P a particular K with $|K| = P$, and then defining $T(P) = T(K)$. Any two such choices give naturally equivalent functors.

6. CONSTRUCTION OF FUNCTORS

Suppose given a contravariant functor $Z : Bi \to S_*$ satisfying the following axioms.

E (extension). Suppose $e : (K_0, L_0) \to (K, L)$ is an expansion; then $Z(e) : Z(K, L) \to Z(K_0, L_0)$ is surjective.

G (glue). Suppose $K = K_1 \cup K_2$, $L \subset K$, and $L_i = L \cap K_i$, $i = 1, 2$. Suppose $u_i \in Z(K_i, L_i)$, $i = 1, 2$, restricts to $u \in Z(K_1 \cap K_2, L_0)$, where $L_0 = L \cap (K_1 \cap K_2)$. Then there exists $z \in Z(K, L)$ so that z restricts to u_i, $i = 1, 2$. Moreover if $K_1 \cap K_2 \subset L$ then z is unique.

Define $T(K, L) = Z(K, L)/\sim$, where $z_0 \sim z_1$ if there is a $z \in Z(K \times I, L \times I)$ so that z_i is identified with the restriction $z| \in Z(K \times \{i\}, L \times \{i\} \cong Z(K, L)$. It is an easy exercise to show that $\sim$ is an equivalence relation. To see that $z \sim z$, consider $Z(K \times \Delta_2, L \times \Delta_2)$.

Now if $f : (K_1, L_1) \to (K_2, L_2)$ is a morphism in Bi, then $T(f) : T(K_2, L_2) \to T(K_1, L_1)$ is clearly well-defined by

$$T(f)[z] = [Z(f)z].$$

Proposition 6.1. The functor T satisfies Axiom C.

Proof. Let $e : (K_0, L_0) \to (K, L)$ be an expansion. We have to show $T(e)$ is an isomorphism. But by Axiom E we have $T(e)$ is onto. Suppose then that $T(e)z_0 = T(e)z_1$. Construct $z \in Z(K \times I, L \times I)$ such that $z : z_0 \sim z_1$ in two steps. First, find $z_2 \in Z(K \cup (K_0 \times I), L \cup (L_0 \times I))$ using Axiom G twice; then find z using E.

Compatibility of extension to Bs

Now suppose that Z is in fact defined on Bs. Then we have T defined using $Z|Bi$, and T extends uniquely to Bs by Theorem 5.1.

Proposition 6.2. The extension of T to Bs is given by

$T(f)[z] = [Z(f)z]$.

Proof. By uniqueness, it suffices to show that $T(f)$ is well-defined on Bs by the above formula, and it is sufficient to consider the case f simplicial. Let K be any cell complex, and define $(K \times I)'$ by deriving each cell on the half-way level. Then if K is simplicial, so is $(K \times I)'$; and if $f : K_1 \to K_2$ is simplicial, then the deriveds may be chosen so that $L(f \times id) : (K_1 \times I)' \to (K_2 \times I)'$ is simplicial. The result therefore follows from

Lemma 6.3. $z_0 \sim z_1$ if and only if there is $z \in Z(K \times I)'$ such that $z_i = Z|K \times \{i\}$.

Proof. Consider $Q = (K \times \Delta^2)'$ obtained by deriving each cell of $K \times \{v_1\}$. Then Q contains isomorphic copies of $(K \times I)'$; namely, $K \times \Delta_1^1$ and $(K \times (\Delta_0^1 \cup \Delta_2^1))'$. Moreover, it is easy to see that Q collapses to both these subsets. Therefore, given a $z \in Z(K \times I)$, we get $\bar{z} \in Z(K \times \Delta^2)'$, and hence $\bar{z}| \in Z(K \times I)'$ and vice versa.

7. COHOMOLOGY THEORIES

Now let $\overline{Bi} \subset Bi$ be the subcategory consisting of pairs (K, L) with $L = \emptyset$, and suppose that $Z : \overline{Bi} \to S_*$ is a functor. Then we can extend Z to Bi by defining $Z(K, L) = \text{Ker}\{Z(K) \to Z(L)\}$. Suppose Z now satisfies axioms E and G. Let T denote the associated homotopy functor.

Lemma 7.1. T <u>satisfies the following axioms for any</u> $K \supset K_1, K_2$:

<u>Half exact:</u> $T(K, K_1 \cup K_2) \to T(K, K_1) \to T(K_2, K_1 \cap K_2)$ is exact.

<u>Excision:</u> $T(K_1 \cup K_2, K_1) \to T(K_2, K_1 \cap K_2)$ is an isomorphism.

Proof. Order 2 is obvious; to see exactness, use E to extend a concordance on K_2 to one on K. For excision, use the definition of Z(K, L) and G.

Now suppose given functors Z^q for $q \in \dot{Z}$ defined on $\overline{Bi}$ and extended to Bi as above, and suppose that in addition we have

<u>Axiom S</u> (suspension).* There are given natural isomorphisms

$$s^q : Z^q(K, L) \to Z^{q-1}(K \times I, K \times \dot{I} \cup L \times I).$$

Then we can define

$$\partial^q : T^q(L) \to T^{q-1}(K, L)$$

to be the composition

$$\begin{array}{ccccc} T^q(L) & \xrightarrow{s^q} & T^{q-1}(L \times I, L \times \dot{I}) & \xleftarrow{T(i)} & T^{q-1}(W, K \cup L \times \{0\}) \\ & \searrow^{\partial^q} & & & \downarrow T(j) \\ & & & & T^{q-1}(K, L) \end{array}$$

where $W = K \cup_{L \times \{1\}} L \times I$, and i is an excision, j extends the identification $L \to L \times \{0\}$, and as a map $K \to W$ is homotopic to the inclusion by an extension of the obvious homotopy on L.

Then easy arguments show that the long sequence is exact, and we have shown

---

* See the note on indexing cohomology groups at the end of the introduction.

Theorem 7.2. A sequence of functors $Z^q : \overline{Bi} \to S_*$, $q \in Z$, satisfying E, G, and S defines a cohomology theory on the category of compact polyhedral pairs.

Remarks 7.3. 1. $T^q(\ )$ is in fact an abelian group functor. This is seen by 'track addition': Given ξ, $\eta \in Z^q(K)$, form $s^q\xi$, $s^q\eta \in Z^{q-1}(K \times I, K \times \dot{I})$, and use G to construct $s^q\xi + s^q\eta \in Z^{q-1}(K \times I)'$. Finally, use amalgamation and inverse of suspension to return to $Z^q(K)$.

2. In fact, half exactness, excision, and suspension imply cohomology theory by formal argument, using Puppe sequences. Thus Axiom S need hold only for $T^q(\ ,\ )$.

3. A classifying Ω-spectrum for $T^q(\ )$ can be constructed by taking a Δ-set $\mathcal{G}_q$ with $\mathcal{G}_q^k = Z^q(\Delta^k)$ then it can be seen that $|\mathcal{G}_q| \simeq \Omega|\mathcal{G}_{q-1}|$. We will explain this construction in detail in Chapter II §5 for a specific example. This extends the theory to infinite complexes.

REFERENCES FOR CHAPTER I

[1] P. Gabriel and M. Zisman. Homotopy theory and calculus of fractions. Springer-Verlag, Berlin (1967).

[2] C. P. Rourke and B. J. Sanderson. Δ-sets I. Quart. J. Math. Oxford Ser. 2 (1971), 321-8.

[3] L. Siebenmann. Infinite simple homotopy types. Proc. Kon. Ned. Akad. (Amsterdam), 73 (1970), 479-95.

[4] J. Stallings. Tata Institute notes on polyhedral topology, 1969.

[5] J. H. C. Whitehead. Combinatorial homotopy I. Bull. Amer. Math. Soc. 55 (1949), 213-45.

[6] E. C. Zeeman. Seminar on combinatorial topology. I. H. E. S. and Warwick notes (1963-69).

[7] S. MacLane. Categories for the working mathematician. Springer-Verlag, New York (1971).

[8] C. P. Rourke and B. J. Sanderson. Introduction to piecewise-linear topology. Springer-Verlag, Berlin (1972).

II·Mock bundles

We describe here the theory of mock bundles. This is a bundle theory giving rise to a cohomology theory which in the simplest case is pl cobordism. In this interpretation, all the usual products have simple definitions, and the Thom isomorphism and duality theorems have short transparent proofs. Another feature is that mock bundles can be composed yielding a cohomology transfer. The theory also provides a short proof of the pl transversality theorem [12; 1.2]. In a final section, classifying Δ-sets are constructed. The construction is similar to Quinn's [7; §1].

1. MOCK BUNDLES AS A COHOMOLOGY THEORY

Let K be a ball complex. A q-mock bundle* ξ^q/K with base K and total space E_ξ consists of a pl projection $p_\xi : E_\xi \to |K|$ such that, for each $\sigma \in K$, $p_\xi^{-1}(\sigma)$ is a compact pl manifold of dimension $q + \dim \sigma$, with boundary $p_\xi^{-1}(\dot{\sigma})$. We denote $p_\xi^{-1}(\sigma)$ by $\xi(\sigma)$, and call it the block over σ.

The empty set is regarded as a manifold of any dimension; thus $\xi(\sigma)$ may be empty for some cells $\sigma \in K$. Therefore, q could be negative, and then $\xi(\sigma) = \emptyset$ if $\dim \sigma < -q$. The empty bundle $\emptyset/K$ has the empty set for total space, and is a q-mock bundle for all $q \in \mathbf{Z}$. Figure 2 shows a 1-mock bundle over the union of two 1-simplexes.

Mock bundles ξ, η/K are isomorphic, written $\xi \cong \eta$, if there is a pl homeomorphism $h : E_\xi \to E_\eta$ which respects blocks; i.e., $h(\xi(\sigma)) = \eta(\sigma)$ for each $\sigma \in K$.

Now define the based set $Z^q(K)$ to be the set of isomorphism classes of q-mock bundles over K with base point the empty bundle.

---

* The terminology 'mock bundle' is due to M. M. Cohen.

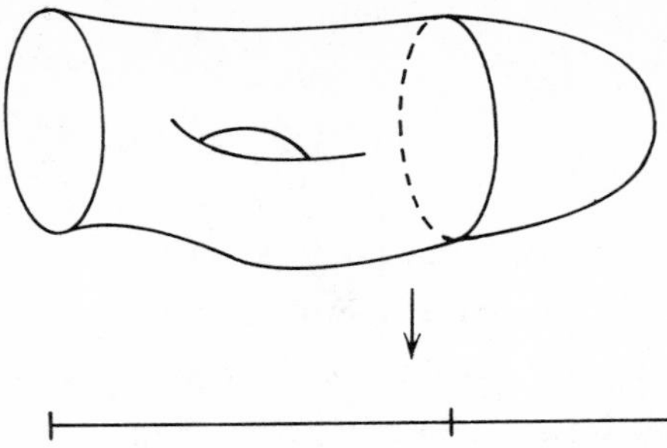

Fig. 2

$Z^q(\ )$ becomes a contravariant functor on the category $\overline{Bi}$ of ball complexes and inclusions via the restriction: Suppose given ξ/K and $L \subset K$; then $\xi|L$ is defined by $E(\xi|L) = p_\xi^{-1}(L)$, and $p(\xi|L) = p_\xi| : E(\xi|L) \to L$. (We use both notations $E(\xi)$ and E_ξ for total spaces, etc., as convenient.)

Now define a functor $Z^q(\ ,\ )$ on the category Bi of pairs of ball complexes (as in I. 7) by defining $Z^q(K, L)$ to be the kernel of $Z^q(K) \to Z^q(L)$. In other words, $Z^q(K, L)$ is isomorphism classes of bundles which are empty over L.

We can now define a functor $T^q(\ ,\ )$, as in I. 6, by taking $T^q(K, L)$ to be the set of cobordism classes of mock bundles empty over L, where ξ_0 is cobordant to ξ_1, written $\xi_0 \sim \xi_1$, if there is a mock bundle $\eta/K \times I$, empty over $L \times I$, such that $\eta|K \times \{i\} \cong \xi_i$ for $i = 0, 1$. It is easy to see that cobordism is an equivalence relation, and in any case we now prove:

Theorem 1. 1. $Z^q(\ ,\ )$ satisfies axioms E, G, and S of Part I, and hence, by I. 7. 2, $\{T^q(\ ,\ ), \partial_q\}$ is a cohomology theory* on the category of pairs of compact polyhedra.

Proof of 1. 1. For Axiom G (glue), suppose given ξ_1/K_1, ξ_2/K_2, and an isomorphism $h : \xi_1|K_1 \cap K_2 \cong \xi_2|K_1 \cap K_2$. Form $\xi/K = K_1 \cup K_2$

* See the note on indexing cohomology groups at the end of the introduction

by $E(\xi) = E(\xi_1) \cup_h E(\xi_2)$, and define the projection of ξ inductively, using the fact that cells are contractible.

For Axiom S (suspension), suppose given ξ/K. Now form $s\xi/K \times I$ by $E(s\xi) = E(\xi)$ and $p_{s\xi} = p_\xi \times \{\frac{1}{2}\}$; i.e., place ξ over $K \times I$ at the half-way level. Then $s\xi$ is empty over $K \times \dot{I}$ (and over $L \times I$ if ξ empty over L), so we have a suspension map

$$s^q : Z^q(K, L) \to Z^{q-1}(K \times I, L \times I \cup K \times \dot{I}).$$

The inverse map is defined by composition with the projection $\pi : K \times I \to K$; i.e., given $\eta/K \times I$ empty over $K \times \dot{I}$, define $s^{-1}\eta/K$ by $E(s^{-1}\eta) = E(\eta)$ and $p(s^{-1}\eta) = \pi \circ p_\eta$. It is trivial to check that s^{-1} is an inverse for s.

Finally, we have to check Axiom E (extension). In other words, if $K \searrow K_0$ and ξ_0/K_0 is given, we have to construct ξ/K so that $\xi_0 \cong \xi|K_0$. By induction on the length of the collapse, it suffices to prove the case when the collapse is elementary across a cell σ from a free face τ. Let J be the subcomplex $\dot{\sigma} - \tau$. Then $|J|$ is a ball, and we can identify $(\sigma, |J|)$ with $(|J| \times I, |J| \times \{0\})$. We then define $E(\xi) = E(\xi_0) \cup E(\xi_0|J) \times I$ identified over $E(\xi_0|J) = E(\xi_0|J) \times \{0\}$, and let $p_\xi = p(\xi_0)$ on $E(\xi_0)$ and $p(\xi_0) \times \mathrm{id}$ on $E(\xi_0|J) \times I$. That ξ is a mock bundle follows from Lemma 1.2 below.

Lemma 1.2. <u>Suppose</u> $|K|$ <u>is a</u> pl n-<u>manifold, and</u> ξ/K <u>is a</u> q-<u>mock bundle. Then</u> E_ξ <u>is an</u> $(n+q)$-<u>manifold with boundary</u> $p_\xi^{-1}(\partial|K|)$.

Proof. (Compare [4; p. 142].) Let $x \in E(\xi)$; then $x \in \operatorname{int} \xi(\sigma)$, say. We can then define a 'transverse star' to x in $E(\xi)$ by inductively restricting collars of $\partial\xi(\tau)$ in $\xi(\tau)$ for $\sigma < \tau$. Then a neighbouthood of x in $E(\xi)$ is homeomorphic to {neighbourhood in $\xi(\sigma)$} $\times$ {transverse star}. The same construction holds for $p(x)$ in K, and the two transverse stars are abstractly isomorphic, hence homeomorphic. But $|K|$ being a manifold implies that the transverse star is a disc ($P \times Q$ is a manifold at (x, y) if and only if P is one at x and Q one at y; see [6]). The result follows on observing that $x \in$ int (transverse star) if and only if $x \in \operatorname{int} |K|$.

The Thom isomorphism

We finish the section by observing that the proof of Axiom S given above generalises at once to give a Thom isomorphism theorem for pl block bundles (see [11] for definitions).

Let u^r/K be a block bundle; then we can give $E(u)$ a ball complex structure in which the blocks of u are balls, for we merely have to choose a suitable ball complex structure on $E(\dot{u})$. Then we can define

$$\Phi : Z^q(K) \to Z^{q-r}(E(u), E(\dot{u}))$$

to be composition with the zero section $i : K \to E(u)$; i.e., $E(\Phi(\xi)) = E(\xi)$ and $p(\Phi(\xi)) = i \circ p_\xi$.

Proposition 1.3. Φ <u>is an isomorphism, and induces an isomorphism, called the Thom isomorphism,</u> $T^q(K) \to T^{q-r}(E(u), E(\dot{u}))$.

Proof. An inverse for Φ is given by composition with a projection for u.

Remark. In the next section, we show that the Thom isomorphism is given by cup product with a Thom class.

2. THE GEOMETRY OF MOCK BUNDLES

We now give geometric interpretations to parts of the cohomology theory $T^q(,)$ defined in §1.

Addition

There is an addition in $Z^q(K, L)$ given by disjoint union; i.e., $E(\xi + \eta) = E(\xi) \cup E(\eta)$ and $p_{\xi+\eta} = p_\xi \cup p_\eta$. It is then easy to check that this coincides with the 'track addition' defined in $T^q(,)$ (I. 7. 3). To see the group structure directly, observe that $\xi + \xi \sim \emptyset$ by letting $\eta/K \times I$ have $E(\eta) = E(\xi \times I)$ and

$$p_\eta(x, t) = \begin{cases} (p_\xi(x), t), & t \le \tfrac{1}{2}, \\ (p_\xi(x), 1 - t), & t \ge \tfrac{1}{2}. \end{cases}$$

Amalgamation

Let ξ^q/K' be given where $K' \lhd K$. Then $p_\xi : E(\xi) \to |K|$ is a mock bundle over K by Lemma 1.2, called the *amalgamation* of ξ and written $am(\xi)$. To see compatibility with amalgamation as in Chapter I §5, notice that the bundle $\eta/K \times I \cup K' \times \{0\}$ obtained by extending $\xi/K' \times \{0\}$ via the proof of 1.1 has total space homeomorphic with $E(\xi) \times I$. This is checked by induction over the skeleta. Uniqueness of collars is used to match product structures.

Subdivision

Now let ξ/K and $K' \lhd K$. Examine the proof of existence of a subdivision ξ'/K' such that $am(\xi') \sim \xi$ (which follows from Chapter I and 1.1). The proof says consider $K'' \lhd K'$ such that K'', as a subdivision of K, is cylindrical in each ball less a smaller ball (see I.2.3). Then $K'' \times \{0\} \cup K \times I \searrow K \times \{1\}$ so that we can extend $\xi/K \times \{1\}$ to $\eta/K'' \times \{0\} \cup K \times I$ and let $\xi' = am(\eta|K'' \times \{0\})$. But the extension is defined skeletally, and, on a typical cell, $\sigma \in K$ is obtained by first extending over the cylindrical collapse $\sigma'' - \sigma_1 \searrow \dot{\sigma}$ and then over the initial collapse $\sigma \times I \searrow \sigma'' - \sigma_1 \cup \dot{\sigma} \times I \cup \sigma \times \{1\}$.

It follows that we can inductively identify $E(\eta)$ with $E(\xi) \times I$ since $E(\eta|\sigma'' - \sigma_1) \cong E(\eta|\sigma) \times I$ can be identified with a collar on $\xi(\sigma)$, and $\eta(\sigma \times I) = \xi(\sigma) \times I$ (by the proof of 1.1). A simple collaring argument is again used to match product structures. We have proved:

Theorem 2.1. *Suppose given* ξ/K *and* $K' \lhd K$. *Then there exists* ξ'/K' *such that* $am(\xi') \cong \xi$.

Remark. The reader can check that this proof of existence of subdivisions is essentially that given in [9; p. 128].

Induced mock bundles

Suppose $f : K \to L$ is a simplicial map between simplicial complexes, and ξ/L a mock bundle. Then we can form the pull-back diagram

$$\begin{array}{ccc} E(f^{\#}(\xi)) & \dashrightarrow & E(\xi) \\ \downarrow & & \downarrow p_\xi \\ K & \xrightarrow{\ f\ } & L \end{array}$$

and it is not hard to show that $f^{\#}(\xi)$ is a mock bundle, and that

$$f^{\#} : Z^q(L) \to Z^q(K)$$

is well-defined and functorial (compare [12; 2.3]). This means that $Z^q(,)$ and, hence by Chapter I, $T^q(,)$ is in fact defined on the larger category Bs, and it follows from I. 6. 2 that the notion of pull-back is compatible with the pull-back as homotopy functor.

The external product

Let ξ^q/K and η^r/L be mock bundles. Then $\xi \times \eta/K \times L$ is defined by $E(\xi \times \eta) = E(\xi) \times E(\eta)$ and $p(\xi \times \eta) = p_\xi \times p_\eta$. The blocks of $\xi \times \eta$ are then products of blocks of ξ and η. We thus get an external product

$$T^q(K) \otimes T^r(L) \to T^{q+r}(K \times L).$$

Remark. In the case that η is the identity $id : L \to L$, we have $[\xi \times \eta] = \pi^*[\xi]$ where $\pi : K \times L \to K$ is the projection. To see this, suppose K, L, and $\pi : (K \times L)' \to K$ are all simplicial. Then $\pi^{\#}(\xi)$ is a subdivision of $\xi \times \eta$.

The internal product

Suppose given ξ^q/K and η^r/K. Define $[\xi] \cup [\eta] \in T^{q+r}(K)$ to be $\Delta^*[\xi \times \eta]$, where $\Delta : K \to K \times K$ is the diagonal map; i. e., subdivide $\xi \times \eta$ so that $\Delta(K)$ is a subcomplex, and then restrict to $\Delta(K)$.

The internal or cup product makes $T^*(K)$ into a commutative ring with unit. To see that the class of $id : K \to K$ is the unit, use the remark on external products and the fact that $\pi \circ \Delta = id$. Associativity and commutativity are easily checked. There are natural relative versions of both products which we leave the reader to formulate.

The composition and the transfer

We now generalise the composition used in §1 to give suspension and Thom isomorphisms.

Let ξ^q/K be a mock bundle (a block bundle with closed manifold as fibre is a special case), and let $\eta^r/E(\xi)$ be another mock bundle.

Now if we subdivide η so that the blocks of ξ are subcomplexes, then corresponding to $\sigma \in K$, we have by 1. 2 the manifold $E(\eta \mid \xi(\sigma))$. In other words, we have a (q+r)-mock bundle

$$p_\xi \circ p_\eta : E(\eta) \to K$$

which we call $\xi \circ \eta$.

$$\begin{array}{ccc} E(\eta) & \xrightarrow{p_\eta} & E(\xi) \\ & \searrow^{p_{\xi \circ \eta}} & \downarrow p_\xi \\ & & K \end{array}$$

This extends to give a transfer

$$p_! : T^r(E(\xi)) \to T^{r+q}(K),$$

where $p = p_\xi$.

We observe that the transfer is functorial i. e. $p_! q_! = (pq)_!$ where p and q are mock bundle projections

$$E(\xi) \xrightarrow{q} E(\eta) \xrightarrow{p} K.$$

The transfer can be seen to be the composition of the following:

$$T^r(E(\xi)) \xrightarrow{\text{Thom iso.}} T^{r-s}(E(\nu), \dot{E}(\nu)) \xrightarrow{i^*} T^{r-s}(K \times I^{q+s}, K \times \dot{I}^{q+1}) \underset{\cong}{\xleftarrow{\text{susp.}}} T^{r+q}(K)$$

where $E(\xi)$ is embedded in $K \times \mathrm{int}(I^{q+s})$ with normal bundle ν^s.

Alternative description of the cup product

Proposition 2. 2. Let $\xi^q, \eta^r/K$ be mock bundles. Then $[\xi] \cup [\eta] = p_! p^*[\xi]$, where $p = p_\eta$.

In other words: pull ξ back over $E(\eta)$ and then compose.

Proof. We may suppose that K is simplicial. Subdivide p so that $p : E(\eta)' \to K'$ is simplicial, and subdivide ξ to ξ'/K'. Let $(K \times K)'$ be the subdivision of $K \times K$ given by [14; Lemma 1] so that $\Delta(K)$ is now a subcomplex. Consider the following diagram.

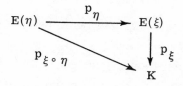

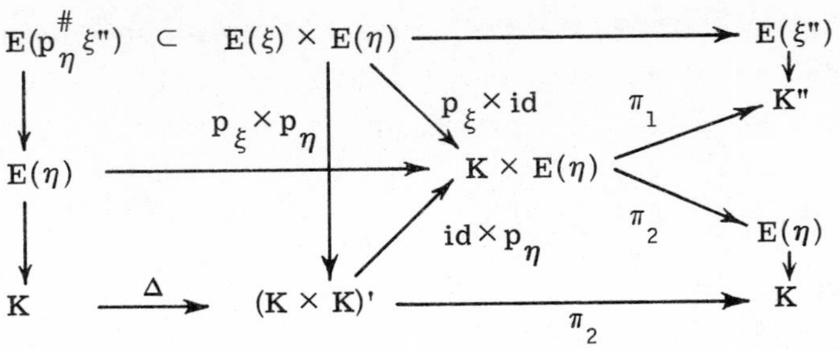

The squares are pull-backs, and K'' is constructed as follows. Choose a triangulation of $K \times E(\eta)$ so that the blocks of $id \times p_\eta$ are subcomplexes. Then choose a further subdivision, and choose $K'' \lhd K'$ so that π_1 is now simplicial. Then choose ξ''/K'' subdividing ξ'/K'. Without loss of generality, we can assume $p_{\xi'} = p_{\xi''}$. Now $p_\xi \times p_\eta$ is the projection of a mock bundle $\zeta/(K \times K)'$, since by construction $\zeta = \pi_2^{\#}(\eta) \circ \pi_1^{\#}(\xi'')$. But from definitions $p_! p^*[\xi] = \Delta^*[\zeta] = \Delta^*[\xi \times \eta] = [\xi] \cup [\eta]$

The Thom class and the Euler class of a block bundle

Let u^r/K be a block bundle with $i : K \to E(u)$ the zero section. If $E(u)$ is given the ball complex structure of §1 (in which blocks are balls), then i is the projection of a mock bundle, and thus determines a class $t(u) \in T^{-r}(E(u), \dot{E}(u))$, called the Thom class of u.

Proposition 2.3. $\cup\, t(u) : T^q(E(u)) \to T^{q-r}(E(u), \dot{E}(u))$ is the restriction isomorphism $i^* : T^q(E(u)) \to T^q(K)$ composed with the Thom isomorphism.

Proof. The result follows from the alternative definition of the cup product, and the fact that the Thom isomorphism is composition with $t(u)$. See diagram.

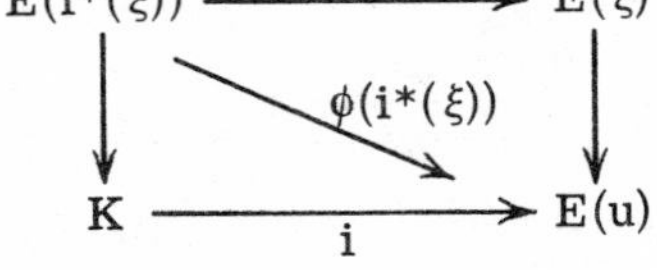

There is also the Euler class of u, $e(u) = i^*t(u)$ which can be thought of as the result of intersecting K with itself in $E(u)$. Both the

Thom and the Euler classes are natural and multiplicative. We leave the reader to check these facts.

Properties of the cup product and the transfer

We now give some properties which will be used in Chapter V.

Proposition 2.4. Suppose given a pull back diagram

$$\begin{array}{ccc} E(f^{\#}\xi) & \xrightarrow{f'} & E(\xi) \\ \downarrow{\scriptstyle p'} & & \downarrow{\scriptstyle p} \\ L & \xrightarrow{f} & K \end{array}$$

with f simplicial. Then $f^*p_! = p'_!f'^*$.

Proof. Subdivide so that $p : E(\xi)' \to K'$ is simplicial, then subdivide so that $f : L' \to K'$ is simplicial. Now $E(f^{\#}\xi)$ is a linear cell complex in a natural way and we may subdivide without introducing new vertices so that p', f' are also simplicial. The result now follows from definitions.

Remark 2.5. In the case $|K| = M$ is a manifold, f is the projection of a mock bundle η/K^+, where $K \lhd K^+$, and f, p are embeddings, then $E(f^{\#}\xi)$ is the transverse intersection of $E(\xi)$ with $E(\eta)$ in M. The proposition then implies that if $g : W \to E(\xi)$ is a map, then we can make g transverse to $E(f^{\#}\xi)$ in $E(\xi)$, or make g transverse to $E(\eta)$ in M. The result in either case is the same. We return to these ideas in §4.

Proposition 2.6. Let p be the projection of a mock bundle η. Then $p_!(p^*\xi \cup \zeta) = \xi \cup p_!\zeta$.

Proof. Consider the diagram:

$$\begin{array}{ccccc} T & \longrightarrow & E(p^*\xi) & \longrightarrow & E(\xi) \\ \downarrow & & \downarrow & & \downarrow{\scriptstyle p_\xi} \\ E(\zeta) & \xrightarrow[p_\zeta]{} & E(\eta) & \xrightarrow[p]{} & K \end{array}$$

(with a dashed arrow from T to K)

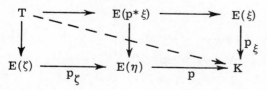

Triangulate $E(\zeta)$, $E(\eta)$, K so blocks are subcomplexes and p_ζ, p are simplicial. Assume $p_\xi = p_{\xi'}$ where $K' \lhd K$ is the resulting subdivision. The result now follows from definitions and 2.2.

3. CAP PRODUCTS AND DUALITY

Let X be a topological space, and define $T_n(X)$ to be the set of <u>bordism</u> classes of pairs (M, f). Here M is a closed pl n-manifold, $f : M \to X$ a continuous map, and (M_0, f_0) is bordant to (M_1, f_1) if there exists $f : W^{n+1} \to X$, where W is a pl manifold with boundary the disjoint union $W_0 \cup W_1$, and there are homeomorphisms $g_i : M_i \to W_i$ such that $f \circ g_i = f_i$, $i = 0, 1$.

$T_n(X)$ becomes a group by 'disjoint union' and is the nth pl bordism group of X.

There are relative groups $T_n(X, A)$ defined by considering maps $f : (M, \partial M) \to (X, A)$ with the notion of bordism enlarged to allow $\partial W = W_0 \cup W_1 \cup W_2$ when W_0 and W_1 are disjoint, and $f(W_2) \subset A$.

There is a homomorphism $\partial_n : T_n(X, A) \to T_{n-1}(A)$ given by restriction. The following theorem is well-known. A sketch proof is included, designed to generalise in Chapters III, IV.

Theorem 3.1. $\{T_n(\ ,\ ); \partial_n\}$ <u>is a homology theory on the category of topological pairs.</u>

Proof. Given a map $f : (X, A) \to (Y, B)$, we get $f_* : T_n(X, A) \to T_n(Y, B)$ by composition. Naturality of f_* and ∂ are then obvious. Exactness is proved by easy geometric arguments. Homotopy follows from the fact that $M \times I$ is an (n+1)-manifold with boundary $M \times \{0\} \cup M \times \{1\} \cup \partial M \times I$. It remains to prove excision. Let $U \subset A$ with $cl(U) \subset int(A)$. Then we will show that

$$i_* : T_n(X - U, A - U) \to T_n(X, A)$$

is an isomorphism.

To see surjectivity, consider $f : (M, \partial M) \to (X, A)$, and define $U_1 = f^{-1}(U)$, $A_1 = f^{-1}(A)$. Then we have $cl(U_1) \subset int_M(A_1)$. We claim that there is a manifold $M_1 \subset M$, with $M - A_1 \subset M_1 \subset M - U_1$. Then

$f_1 = f|M_1$ defines an element of $T_n(X - U, A - U)$, and $f \circ \pi_1 : M \times I = W \to X$ defines a bordism between f and f_1 where $W_0 = M \times \{0\}$, $W_1 = M_1 \times \{1\}$. To construct M_1, give M a metric, and let $\varepsilon = d(M - \text{int}(A_1), cl(U_1))$, and triangulate M with mesh $< \varepsilon/2$. Define $P = \cup$ of closed simplexes which meet $M - A_1$, and let M_1 = 2nd derived neighbourhood of P (cf. [14]). The required properties are then easily checked. Injectivity follows by a similar argument applied to a bordism.

The cap product

Let $f : M^n \to K$ be a map, where M is a closed pl n-manifold, and let ξ^q/K be a mock bundle. Form $f^*[\xi]/M$, and choose a representative η. Then by 1.2, $E(\eta)$ is a closed (n+q)-manifold, and there is a map $g : E(\eta) \to |K|$ by composing. Notice that if f is simplicial, then we can take $\eta = f^{\#}(\xi)$.

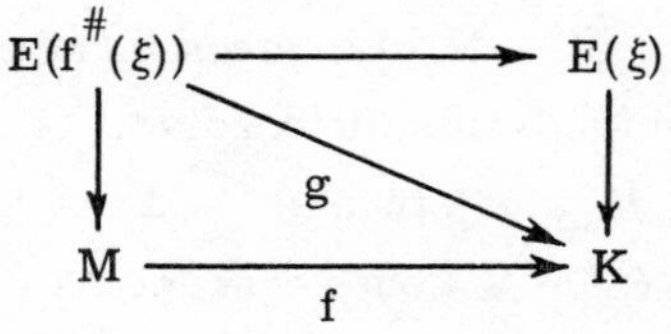

We define $[f] \cap [\xi] = [g]$, and the reader may check that we have a well-defined cap product

$$\cap : T_n(|K|) \otimes T^q(K) \to T_{n+q}(|K|).$$

Remarks. 1. There are obvious relative versions of the cap products.

2. The similarity of the cup and cap products (using the second description of the cup product) is clear. This will become more transparent later when they are seen to be dual.

Slant products

1. $T_q(X) \otimes T^s(X \times Y) \to T^{q+s}(Y)$

$[f] \quad \otimes \quad [\xi] \quad \to \pi \circ (f \times 1_Y)^*[\xi]$.

Consider $M \times Y \xrightarrow{f \times 1_Y} X \times Y \to Y$. Then we have the bundle (with $\pi : M \times Y \to Y$ the composite)

$(f \times 1_Y)^* \xi/M \times Y$, which we can regard as a bundle over Y by composing with π.

2. $$T^q(X) \otimes T_s(X \times Y) \to T_{q+s}(Y)$$
$$[\xi] \otimes [f] \mapsto \pi_Y \circ([f] \cap [\xi \times 1_Y]).$$

In other words, take $\xi \times 1_Y$ as a bundle over $X \times Y$, take its cap product with f, and then compose into Y.

Remark. As before, there are relative versions which give in particular slant products when X is replaced by (X, x_0), Y by (Y, y_0), and $X \times Y$ by $X \wedge Y \equiv (X \times Y, X \vee Y)$.

Poincaré duality

Let M be a closed n-manifold and ξ/M a q-mock bundle over some complex underlying M. Now the fundamental (bordism) class $[M]$ of M is just the identity map $1 : M \to M$, and since $1^* = 1$, $[M] \cap \xi$ can be interpreted as the same map $p_\xi : E(\xi) \to M$, but regarded as a bordism class by 1.2. In other words, we have the Poincaré duality map

$$\psi : T^q(M) \to T_{n+q}(M)$$

defined by $\psi\{p : E \to M\} = \{p : E \to M\}$!

Theorem 3.2. ψ <u>is an isomorphism.</u>

Proof. ψ <u>is onto:</u> Let $f : W \to M$ be a bordism class. We have to find a mock bundle which 'amalgamates' to W. We can suppose f is simplicial, and consider the dual cell complex M^* to M. Then $f : W \to M^*$ is the projection of a mock bundle. This is a consequence of Cohen's [3; 5.6]. Notice that for $\alpha^* \in M^*$, we have the block over α^* equal to $D(\alpha, f)$, which is a manifold with boundary corresponding to the boundary of α^*.

ψ <u>is</u> 1 : 1. Suppose ξ^q, η^q/K, $|K| = M$ are mock bundles, and that $f : W \to M$ is a map where W is an (n+q+1)-manifold with boundary the disjoint union $E(\xi) \cup E(\eta)$. We have to show that $\xi \sim \eta$ in $T^q(K)$. Using collars on the boundaries of W, we can replace f by a map $f_1 : W \to M \times I$ so that $f_1^{-1}(M \times 0) = \xi$ and $f_1^{-1}(M \times 1) = \eta$. Subdivide f_1 so that it is simplicial, and so that the blocks of ξ and η and the cells of $K \times 0 \cup K \times 1$ are all subcomplexes. Now consider the mock bundle $f_1 : E(\zeta) \to J$ where J is the cell complex which has for cells the duals in $M \times I$ to simplexes $\alpha \in M \times I$ and the duals in each cell $\sigma \in K \times (1 \cup 0)$ of simplexes $\alpha \in \sigma$. That ζ is a mock bundle follows from [3]. Notice that $\zeta | (K \times 0)'$ is a subdivision of ξ, and similarly for $\zeta | (K \times 1)'$. Now subdivide ζ so that cells $\sigma \times I$ are subcomplexes of its base, for $\sigma \in K$. Finally, amalgamate over $K \times I$ to realise the required cobordism.

Remark. The proof of duality shows that there is really no distinction between bordism and cobordism classes when the base is a manifold; they are represented by precisely the same class of map! The duality between cap and cup products is also clear from the definitions using the duality map; i. e., $\psi(\xi) \cap \eta = \psi(\xi \cup \eta)$, etc. See the next section for connections with transversatility.

General duality theorem

Now let $Y \subset X \subset M$ be compact subpolyhedra, and denote X^c, Y^c for their complements. We can regard X^c, Y^c also as compact polyhedra by removing the interior of derived neighbourhoods of X and Y.

Duality Theorem 3. 3. <u>There is a natural isomorphism</u>

$$\phi : T^q(X, Y) \to T_{n+q}(Y^c, X^c).$$

Remarks. 1. ϕ can be regarded as an extension of the cap product with [M].

2. (3. 3) generalises both Lefshetz duality and Spanier-Whitehead duality; e. g. , for the latter, take $X = M = S^n$ (cf. Whitehead [13]).

3. By Spanier-Whitehead duality, $T^q(\ ,\ )$ is indeed the dual theory to pl bordism.

Proof of 3.3. Let $N(X)$, $N(Y)$ be derived neighbourhoods of X and Y, and define ϕ to be the composition

$$\begin{array}{ccc} T^q(X, Y) \underset{i^*}{\overset{\simeq}{\leftarrow}} T^q(N(X), N(Y)) & \overset{a}{\rightarrow} & T_{n+q}(N(X) - N(Y),\ \dot{N}(X) - \dot{N}(Y)) \\ & \searrow^{\phi} & \downarrow i_* \\ & & T_{n+q}(Y^c, X^c) \end{array}$$

where a is amalgamation; see Fig. 3.

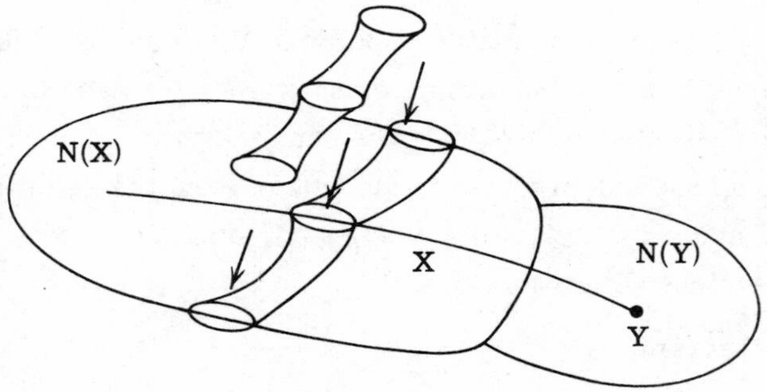

Fig. 3

To see that ϕ is surjective, regard $f : M \rightarrow Y^c$ as the projection of a mock bundle (in which the blocks might have extra boundary over X^c) by Cohen's theorem. Then restrict to X to get a genuine mock bundle. To see ϕ is injective, combine this proof with the second half of the proof of 3.2.

4. APPLICATION TO TRANSVERSALITY

We observe that the mock bundle subdivision theorem (together with Cohen [3; 5.6]) implies various transversality theorems. We deal with the simplest case first, the case of making a map transverse to a submanifold. Then we extend to give relative theorems, theorems for embeddings and for general subpolyhedra. We also give the connection with block transversality [12].

Let $f : W \to M$ be a map between compact $p\ell$ manifolds with W closed, and suppose that $N \subset M$ is a submanifold. Then we can regard $f : W \to M^*$ as the projection of a mock bundle ξ, as in §3. Now ξ can be subdivided so that N is a subcomplex of the base (this involves a homotopy of f); then $f^{-1}(N)$ is the restriction of ξ to N, and hence a manifold by 1.2, and f is now transverse (in some sense) to N!!

Notice that the proof is easily adapted to give an ε-version by making the diameters of cells of $M^* < \varepsilon$. Further, N can be replaced by a whole family of manifolds. In fact, the natural setting is where N is a general subpolyhedron. We now show how to treat relative transverslity in this setting. Let $X \subset M$ be a subpolyhedron. We say that $f : W \to M$ is <u>mock transverse</u> to X if f is the projection of a mock bundle in which X underlies some subcomplex of the base. We write $W \pitchfork_m X$, or $f \pitchfork_m X$.

For technical reasons, we need a condition on $X_0 = X \cap \partial M$ to get a relative theorem (there are counterexamples otherwise; see 4.2 and 4.3 below). We say that X_0 is <u>locally collared</u> in (M, X) provided that at each point $x \in X_0$ there is a neighbourhood in (M, X) which is the product of a neighbourhood in $(\partial M, X_0)$ with the unit interval. Local collaring is equivalent to collaring [8; p. 321].

Relative transversality theorem 4.1. <u>Let M be a compact manifold with boundary and $X \subset M$ a polyhedron with $X_0 = X \cap \partial M$ locally collared in (M, X). Let $f : W \to M$ be a map such that $f^{-1}\partial M = \partial W$, and suppose $f|\partial W \pitchfork_m X_0$; then there is an ε-homotopy of f rel ∂W making f mock transverse to X.</u>

Proof. Suppose $f|\partial W$ is the projection of the mock bundle ξ/K, and choose a ball complex L with $|L| = M$ extending K, and so that X is a subcomplex of L. This is done by first extending to a collar via the product ball complex $K \times I$, and then choosing any suitable ball structure on $M - |K| \times [0, 1)$ and adjoining the two. Following the proof of 3.2, we can suppose that f is the projection of a mock bundle ζ such that

$\zeta | \partial M$ is a subdivision of ξ. Choose a further subdivision ζ'/L' of ζ so that $L' \lhd L$. Then amalgamating over L, we have ξ_1, say, over L with $\xi_1 | K \cong \xi$. It only remains to observe that the homotopy of p_ξ takes place within cells of K, and hence can be shrunk to the identity, and this extends to give a modified homotopy of f by the HEP.

Remark 4.2. In fact, the proof of 4.1 used only that K extends to L with X a subcomplex of L. This needs a much weaker condition on X_0 than local collaring. A necessary and sufficient condition is that the ambient intrinsic dimension [1] of X at X_0 is constant on the interiors of balls of K. This is always true if the ambient intrinsic dimension of X at $x \in X_0$ equals that of X_0 at x.

Example 4.3. Let X = 'the letter T' with the top in ∂M so that $K \cap X$ is a 1-cell. Then, if $f : \partial W \to \partial M$ is not transverse to the mid-point of $K \cap X$, there is no homotopy of f rel ∂W making f mock transverse to X.

Transversality for embeddings

Now suppose that $f : W \to M$ is a locally flat embedding (the condition on local-flatness will be removed later). We say that $f : W \to M$ is an embedded mock bundle if f is the projection of a mock bundle ξ/K, and for each ball $\sigma \in K$, we have $f| : \xi(\sigma) \to \sigma$ is a proper locally-flat embedding (i. e., $f^{-1}(\dot{\sigma}) = \partial\xi(\sigma)$, and f looks locally like the inclusion $R^k_+ \subset R^n_+$ for some k, n). We then observe that the subdivision theorem for mock bundles applies to embedded mock bundles to yield an embedded mock bundle, and that the homotopy which takes place in the proof can be replaced by an ambient isotopy (by uniqueness of collars [5]). Thus Theorem 4.1 applies to give a relative transverality theorem for embeddings via an ε-ambient isotopy (fixed on ∂M).

Connection with block transversality

Suppose $X \subset M$ is a compact polyhedron and W is a locally flat submanifold of M. Recall [12] that X is block transverse to W in M if there is a normal block bundle ν/W in M such that $X \cap E(\nu) = E(\nu | X \cap W)$. We write $X \underline{\pitchfork}_b W$ or $X \underline{\pitchfork}_b \nu$.

Theorem 4.4. $X \pitchfork_b W$ <u>if and only if</u> $W \pitchfork_m X$.

Thus Theorem 4.1 (the version for embeddings) recovers a strengthened form of [12; 1.2].

Proof. Suppose $X \pitchfork_b W$; then there is a normal block bundle ν/W so that $X \cap E(\nu)$ is a union of blocks. Choose a ball complex K with $|K| = M$ so that the blocks of ν are balls of K, and so that X underlies a subcomplex (i.e., simply triangulate the complement of $E(\nu)$ and throw in the blocks of ν!). Then the inclusion $W \subset |K|$ is the projection of an (embedded) mock bundle with X a subcomplex of the base. Notice that the restriction of this mock bundle to $E(\nu)$ gives the Thom class of ν.

Conversely, suppose $W \pitchfork_m X$ by ξ/K; then we construct a normal block bundle ν on W in M by induction over the skeleta of K so that it restricts to a normal bundle for $\xi(\sigma)$ in σ for each $\sigma \in K$. This is an easy consequence of the relative existence theorem for block bundles [11; 4.3]. Then $X \pitchfork_b \nu$.

Extension to polyhedra

This subsection anticipates §3 of Chapter III (manifolds with singularities). Observe that if $f : Y \to M$ is a map where Y is a polyhedron, and we apply the process of Cohen's theorem [3; 5.6] and regard f as a 'bundle' over M^*, then Cohen's proof shows that the blocks, although not manifolds, are polyhedra with collarable 'boundaries'; i.e., if $\sigma \in M^*$, then $f^{-1}(\dot{\sigma})$ is collarable in $f^{-1}(\sigma)$. So we define $f : Y \to K$ to be a <u>polyhedral mock bundle</u> if for each $\sigma \in K$ we have $f^{-1}(\dot{\sigma})$ collarable in $f^{-1}(\sigma)$. Then the subdivision theorem works and we thus get a transversality theorem for two polyhedra in a manifold and similar relative versions and versions for embeddings. In the case of embeddings, mock transversality implies transversality in the sense of Armstrong [2]. This is proved by using the collars to construct neighbourhoods of the form cone $\times$ transverse star; compare with the proof of 1.2. McCrory [15] has shown that mock transversality for polyhedra is symmetric and equivalent to both block transversality in the sense of Stone [16] and to transimpliciality in the sense of Armstrong [2].

The transversality definition of the cup product

Suppose given ξ^q, η^r/K then for some large m we may assume $E(\xi), E(\eta) \subset |K| \times I^m$ so that p_ξ, p_η are restrictions of the projection $|K| \times I^m \to |K|$. Now consider $E(\xi) \times I_0^m, E(\eta) \times I_1^m \subset |K| \times I_0^m \times I_1^m$. By inductively making $\xi(\sigma) \times I_0^m$ transverse to $\eta(\sigma) \times I_1^m$ in $\sigma \times I_0^m \times I_1^m$ we get a mock bundle $E(\xi) \times I_0^m \cap E(\eta) \times I_1^m \to K$. It can easily be seen that this gives the cup product, using the alternative definition. However we sketch a proof below connecting the transversality definition with the restriction to the diagonal definition. This proof has the virtue of generalising to the more complex situation considered in Chapter V.

Let s_q denote q-fold suspension. From definitions $s_{2m}(\xi \times \eta) = s_m(\xi) \times t(K \times I^m) \cup t(K \times I^m) \times s_m(\eta)$ and the total spaces on the right are transverse without being moved. Let $i : |K| \times I^{2m} \to |K| \times |K| \times I^{2m}$ be given by $i(x, y) = (x, x, y)$. Then applying i^* we get

$$s_{2m}(\xi \cup \eta) = s_m(\xi) \times t(I^m) \cup t(I^m) \times s_m(\eta).$$

Desuspending both sides reveals the coincidence of definitions.

5. THE CLASSIFYING SPECTRUM

Ω-spectra and Δ-sets

We give a simple-minded definition of spectra in the category of Δ-sets. Basic facts about Δ-sets contained in [10] will be assumed. Given a based Kan Δ-set X, we define a based Kan Δ-set ΩX as follows.

$$(\Omega X)^{(n)} \subset X^{(n+1)} \quad \text{and} \quad \sigma \in (\Omega X)^{(n)}$$

if and only if $\partial_{n+1}\sigma = *_n$ and $\partial_0^{n+1}\sigma = *_0$. The operators $\partial_i : (\Omega X)^{(n)} \to (\Omega X)^{(n-1)}$, $i = 0, 1, \ldots, n$, are just restrictions of the ∂_i in X.

Let ΩY denote the space of loops on the based CW complex Y, and let SY denote the singular complex of Y. There is an identification

$\Delta^n \times I/\Delta^n \times \{0\} \to \Delta^{n+1}$ given by

$$(t_0, \ldots, t_n, s) \mapsto ((1 - s)t_0, \ldots, (1 - s)t_n, s).$$

This determines a natural isomorphism $\theta(Y) : \Omega SY \to S\Omega Y$. There are also based homotopy equivalences

$$\phi(Y) : |SY| \to Y \quad \text{and} \quad \psi(X) : X \to S|X|$$

(see [10; p. 334]). Consider the composition

$$\phi(\Omega|X|) \circ |\theta|X|| \circ |\Omega\psi(X)| : |\Omega X| \to |\Omega S|X|| \to |S\Omega|X|| \to \Omega|X|.$$

We have now proved:

Lemma 5.1. Suppose X is a Kan based Δ-set. There is a based (weak) homotopy equivalence

$$|\Omega X| \to \Omega|X|.$$

Now define an Ω-spectrum as follows. For each $m \in \mathbf{Z}$ is given a Kan Δ-set $\mathcal{G}_m$ and a homotopy equivalence

$$e_m : \mathcal{G}_m \to \Omega\mathcal{G}_{m-1}.$$

It follows from 5.1 and [13] that a cohomology theory h^* is defined on the category of pairs of CW complexes and homotopy classes of maps by

$$h^m(A, B) = [(A, B), (|\mathcal{G}_m|, |*|)].$$

The Ω-spectra for pl cobordism

Define $\mathcal{G}(PL)_m$ as follows. Let $\mathbf{R}^\infty = \cup \mathbf{R}^n$. Then a k-simplex of $\mathcal{G}(PL)_m$ is a compact polyhedron $X \subset \Delta^k \times \mathbf{R}^\infty$ such that $\pi| : X \to \Delta^k$ is the projection of an m-mock bundle over Δ^k, where π is the natural projection.

Base simplexes $*_k \in \mathcal{G}(PL)_m$ are defined by taking $X = \emptyset$, and face operators are defined by restriction. It follows from the proof of 1.1 and general position that $\mathcal{G}(PL)_m$ is a Kan complex (compare [11;

2.3]). Define $e^k : \Delta^k \to \Delta^{k+1}$ by $e^k(s) = \frac{1}{2}s + \frac{1}{2}v_{k+1}$. Then we have $e^k_m : \mathcal{G}(PL)^{(k)} \to (\Omega\mathcal{G}(PL)_{m-1})^{(k)}$ defined by

$$X \subset \Delta_k \times \mathbf{R}^\infty \xrightarrow{e^k \times id} \Delta_{k+1} \times \mathbf{R}^\infty.$$

$e_m : \mathcal{G}(PL)_m \to \Omega\mathcal{G}(PL)_{m-1}$ is then a based Δ-map.

Proposition 5.2. $e_m : \mathcal{G}(PL)_m \to \Omega\mathcal{G}(PL)_{m-1}$ is a homotopy equivalence.

Proof. e_m is injective so we have to describe a deformation retraction of $\Omega\mathcal{G}(PL)_{m-1}$ on $e_m(\mathcal{G}(PL)_m)$. This can be thought of as 'sliding to the half-way level'. More precisely, suppose given a map $\Lambda_{n,i} \to \Omega\mathcal{G}(PL)_{m-1}$ whose boundary goes into $e_m(\mathcal{G}(PL)_m)$. Then glue the simplexes together to form a mock bundle $\xi/C\Lambda_{n,i}$ embedded in $C\Lambda_{n,i} \times \mathbf{R}^\infty$ and empty over cone-point and base, where $C\Lambda_{n,i}$ denotes a cone on $\Lambda_{n,i}$ with cone-point last. Then $E(\xi)$ lies over the half-way level in $C\partial\Lambda_{n,i}$, and we homotope $E(\xi)$ rel boundary into the pre-image of the half-way level over $C\Lambda_{n,i}$. Then, using an identification of $C\Delta^n$ with $C\Lambda_{n,i} \times I$ and general position, we get an n-simplex of $\Omega\mathcal{G}(PL)_m$ whose restriction to the ith face lies in $e_m(\mathcal{G}(PL)_m)$. The deformation then follows from [10; 6.3].

Now let K be an ordered simplicial complex, and $f : K \to \mathcal{G}(PL)_m$ a Δ-map. Then we can form an m-mock bundle ξ/K by gluing together the images of simplexes of K, and since the base complex gives the empty bundle, we get a function

$$\Phi : [K, L; \mathcal{G}(PL)_m, *]_\Delta \to T^m(K, L),$$

where $[\ ]_\Delta$ denote Δ-homotopy classes. Then by general position (compare [11; § 2]), Φ is a bijection, and it follows from [10] that Φ induces a natural bijection

$$[P, Q; |\mathcal{G}(PL)_m|, |*|] \to T^m(P, Q)$$

for polyhedral pairs.

Finally, we notice that the suspension isomorphism essentially coincides with the function e_m. More precisely, $e_m \circ f : K \to \Omega(\mathcal{G}(PL)_{m-1})$

gives a mock bundle η/CK such that $s\xi/K \times I$ is isomorphic to the amalgamation of the pull-back $\pi^{\#}\eta$ by the pinching map $\pi : (K \times I)' \to K$, where $(K \times I)' \lhd K \times I$ and π is simplicial.

We have proved:

Theorem 5.3. The cohomology theory $\{T^q, \partial_q\}$ coincides for polyhedral pairs with the cohomology theory defined by $\{\mathcal{G}(PL)_m, e_m\}$.

REFERENCES FOR CHAPTER II

[1] E. Akin. Manifold phenomena in the theory of polyhedra. Trans. A. M. S., 143 (1969), 413-73.

[2] M. A. Armstrong. Transversality for polyhedra. Ann. of Math. 86 (1967), 172-91.

[3] M. M. Cohen. Simplicial structures and transverse cellularity. Ann. of Math. 85 (1967), 218-45.

[4] M. M. Cohen and D. P. Sullivan. On the regular neighbourhood of a two-sided submanifold. Topology, 9 (1970), 141-7.

[5] J. F. P. Hudson and E. C. Zeeman. On combinatorial isotopy. Publ. I. H. E. S., 19 (1964), 69-94.

[6] H. Morton. Joins of polyhedra. Topology, 9 (1970), 243-9.

[7] F. S. Quinn. A geometric formulation of surgery. Ph. D. thesis, Princeton University, 1969.

[8] C. P. Rourke. Covering the track of an isotopy. Proc. Amer. Math. Soc., 18 (1967), 320-4.

[9] C. P. Rourke. Block structures in geometric and algebraic topology. Report to I. C. M., Nice, 1970.

[10] C. P. Rourke and B. J. Sanderson. Δ-sets I. Quart. J. Math. Oxford, 2, 22 (1971), 321-8.

[11] C. P. Rourke and B. J. Sanderson. Block bundles I. Ann. of Math., 87 (1968), 1-28.

[12] C. P. Rourke and B. J. Sanderson. Block bundles II. Ann. of Math., 87 (1968), 256-78.

[13] G. W. Whitehead. Generalised homology theories. Trans. A. M. S. 102 (1962), 227-83.

[14] E. C. Zeeman. Seminar on combinatorial topology. I. H. E. S. and Warwick, 1963-66.

[15] C. McCrory. Cone complexes and pl transversality. Trans. A. M. S. 207 (1975), 1-23.

[16] D. Stone. Stratified polyhedra. Springer-Verlag lecture notes, no. 252.

III · Coefficients

In this chapter we give a geometric treatment of coefficients in oriented pl bordism theory. The definition (although not all the theorems) extend to other geometric homology theories and this extension will be covered in Chapter IV, as will the extension to cobordism (mock bundle) theories. Application to general homology theories and connection with other definitions of coefficients will be covered in Chapter VII.

There are two good definitions of coefficients:

1. For a short resolution ρ of an abelian group G we define coefficients in ρ by labelling with generators and introducing one stratum of singularities of codimension 1 corresponding to the relations (see §1).

2. We allow labelling by any group elements, and singularities corresponding to any relation and then, in the bordisms, allow singularities of codimension 2 corresponding to 'relations between relations' (see §3).

Definition 1 is very simple geometrically while definition 2 is functorial in G. To prove equivalence of the two definitions involves a further definition, for longer resolutions (in §2). The basic geometrical trick is resolution of singularities and appears in the proof of the universal coefficient sequence in §2. The universal coefficient sequence itself can be seen as the measure of the obstruction to resolution of the final singularity. In §3 it is seen that the universal coefficient sequence is natural for G; consequently by [3] it splits for a large class of abelian groups, including all groups of finite type.

In §4 we show how to regard the product of a (G, i)-manifold with a (G', i')-manifold as a ($G\otimes G'$, $i+i'$)-manifold and thus define a cross product for bordism with coefficients.

In §5 is the Bockstein sequence and in §6 we observe that, if G is an R-module, then $\Omega_*(-, -; G)$ inherits an $\Omega_*(\text{pt.}; R)$-module structure in a natural way. Using Dold [1] we then have a spectral sequence $\text{Tor}_p(\Omega_q(-, -; R), G) \Rightarrow \Omega_*(-, -; G)$.

We use the convention $- \longrightarrow +$ for inducing orientations on the boundary of 1-manifolds, and in general use the 'inward normal last' convention.

1. COEFFICIENTS IN A SHORT RESOLUTION

Let F be a free abelian group with basis B. We define coefficients in F by labelling with elements of B, i.e. define an F-manifold to be an (oriented pl) manifold each component of which is labelled by an element of B. We write (M, b) or $M \otimes b$ for M labelled by b. It is easy to see that bordism of F-manifolds defines a bordism theory $\Omega_*(, ; F)$ and that $\Omega_*(X, A; F) \cong \Omega_*(X, A) \otimes F$ for the pair (X, A).

Now let G be a general abelian group and ρ a <u>short resolution</u> of G. ρ comprises a short exact sequence

$$0 \to F_1 \xrightarrow{\phi_1} F_0 \xrightarrow{\varepsilon} G \to 0$$

where F_0 and F_1 are free abelian groups with bases B_0 and B_1. The elements of B_0 are the <u>generators</u> of G and those of B_1 are the <u>relations</u> for G with respect to the resolution ρ. We will define coefficients in ρ by starting with $\Omega_*(, ; F_0)$ and 'killing' the elements of $\Omega_0(\text{pt.}; F_0)$ which correspond to the relations:-

Let $r \in B_1$ then $\phi_1(r) \in F_0$ and can be written uniquely in the form $\sum \alpha_i b_i$ where α_i is an integer and the sum is taken over elements $b_i \in B_0$. The element $L(r, \rho) \in \Omega_0(\text{pt.}; F_0)$ is the union of $|\alpha_i|$ points labelled by b_i and oriented + if $\alpha_i > 0$ and - if $\alpha_i < 0$, where the union is taken over all elements $b_i \in B_0$.

A ρ-<u>manifold</u> of dimension n is a polyhedron P with two strata $P \supset S(P)$, labellings and extra structure such that

1. $P - S(P)$ is an F_0-manifold of dimension n.
2. $S(P)$ is an F_1-manifold of dimension $(n - 1)$.

3. For each component (Q, r) of $S(P)$ there is given a regular neighbourhood N of Q in P and an isomorphism

$$h : N \to Q \times C(L(r, \rho))$$

where $C(X)$ denotes the cone on X and h carries Q by the identity to $Q \times$ (cone point).

4. h is an isomorphism of F_0-manifolds off Q.

Intuitively, a ρ-manifold is a manifold labelled by generators for G with codimension 1 singularity labelled by relations. Moreover the sheets and orientations of P near $S(P)$ give the relation labelling the singularity.

Examples 1.1. 1. $G = F$ a free abelian group and $F_1 = 0$, $F_0 = F$, $B_0 = B$. Then a ρ-manifold is precisely an F-manifold.

2. $G = \mathbf{Z}_n$ and we use the usual presentation

$$0 \to \mathbf{Z} \xrightarrow{\times n} \mathbf{Z} \to \mathbf{Z}_n \to 0.$$

Then $B_0 = \{1\}$, $B_1 = \{1\}$ and so the labels give no information and can be suppressed. $L(1, \rho)$ is the union of n points (all oriented $+$). Thus a ρ-manifold is a manifold with a codimension 1 singularity, which has a trivial neighbourhood of the form C(n points). Moreover the orientations of the n sheets all induce the same orientation on the singulatiry. This is precisely Sullivan's description of '$\mathbf{Z}_n$-manifold' [4]. See Fig. 4.

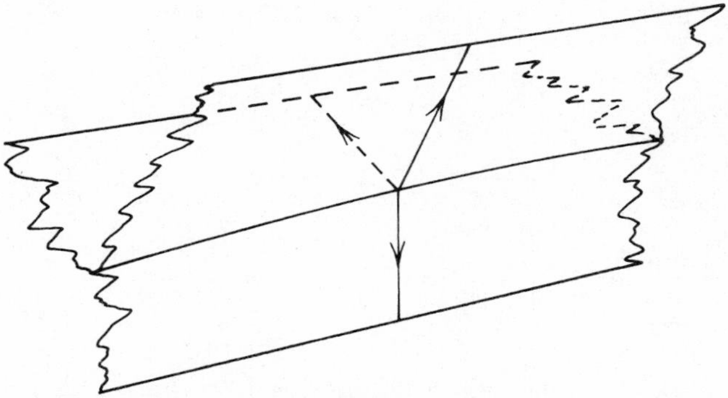

Fig. 4. Part of the neighbourhood of the singularity in a '$\mathbf{Z}_3$-manifold'

3. $G = \mathbf{Z}_{p^\infty}$. We describe the resolution by specifying the generators and relations. The generators are '$1/p$', '$1/p^2$', ... and the relations, as elements of F_0, are $(p('1/p^i') - '1/p^{i-1}')$, $i = 1, 2 \ldots$ and $p('1/p')$. Thus the sheets are labelled '$1/p^i$' and the singularities occur where p sheets labelled '$1/p^i$' merge into one sheet labelled '$1/p^{i-1}$' or where p sheets labelled '$1/p$' merge together. Orientation has the obvious compatibility. See Fig. 5, in which $p = 5$:

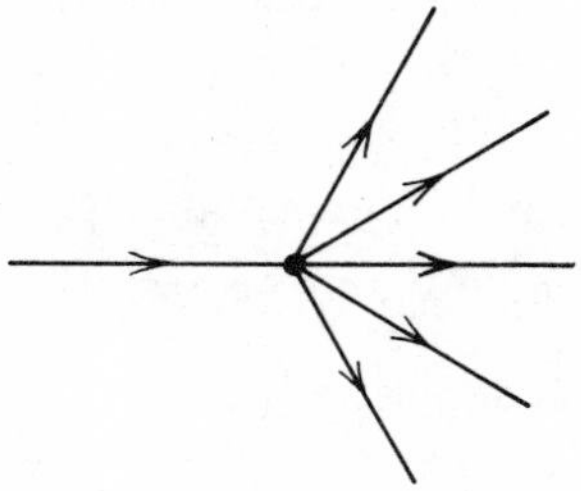

Fig. 5

4. $G = Q$, the rationals. Generators '$1/n$', $n = 1, 2, \ldots$ and relations $p('1/n') - '1/q'$, where $n = pq$ and p is the smallest prime occurring in n. The picture is similar to the last one.

5. In all the cases above we have chosen the most natural resolution for G. Other resolutions also give rise to a notion of 'G-manifold'. In the example below, a, b, c, d are the generators of the copies of $\mathbf{Z}$ indicated:

$$\mathbf{Z} \oplus \mathbf{Z} \longrightarrow \mathbf{Z} \oplus \mathbf{Z} \longrightarrow \mathbf{Z}_3$$

$$b \longmapsto 1$$

$$a \longmapsto 2$$

$$c \longmapsto a - 2b$$

$$d \longmapsto a + b$$

Here a ρ-manifold has two sorts of sheet: a-sheet and b-sheet. Two b-sheets can merge into an a-sheet and an a and b sheet can merge together. This notion gives the same bordism theory as the notion of '$\mathbf{Z}_3$-manifold' in Example 2, as we will show in §3.

There is a natural notion of ρ-manifold with boundary and we thus have a bordism theory $\Omega_*(\, , \, ; \rho)$. That this theory is a generalised homology theory follows from the proof given in II 3.1 for $T_*(\, ,)$. (The crucial fact is that the regular neighbourhood of a polyhedron in a ρ-manifold can be given an essentially unique structure as a ρ-manifold.) We now turn to the universal coefficient theorem for ρ-bordism. In §2 we will prove the theorem in the general case (for longer resolutions) and here we will content ourselves with the statement and a sketch proof of this (simpler) case, with stress on the geometry of the situation.

Theorem 1.2. Let ρ be a short resolution. There is a short exact sequence, natural in (X, A):

$$0 \to \Omega_n(X, A) \otimes G \xrightarrow{l} \Omega_n(X, A; \rho) \xrightarrow{s} \mathrm{Tor}(\Omega_{n-1}(X, A), G) \to 0.$$

Sketch of proof. (For full proof see 2.5.) The spaces (X, A) play no role in the proof of the theorem and we will ignore them.

The notation is intended to suggest that l is the 'labelling' map and s is 'restriction to the singularity'.

Description of l

Using the description of $\Omega_n \otimes F_0$ as manifolds labelled by elements of B_0 (see start of this section) we have a 'labelling' map $l_1 : \Omega_n \otimes F_0 \to \Omega_n(\rho)$. Now l_1 is zero on relations. This is seen as follows. Let r be a relation and $[M] \in \Omega_n$. Consider the labelled manifold $M \times L(r, \rho) \times I$ with each copy of $M \times \{0\}$ identified together and this end labelled by r. This constructs a ρ-bordism of $l_1([M], r)$ to zero. See Fig. 6. It follows that l_1 defines a monomorphism $l : \Omega_n \otimes G \to \Omega_n(\rho)$.

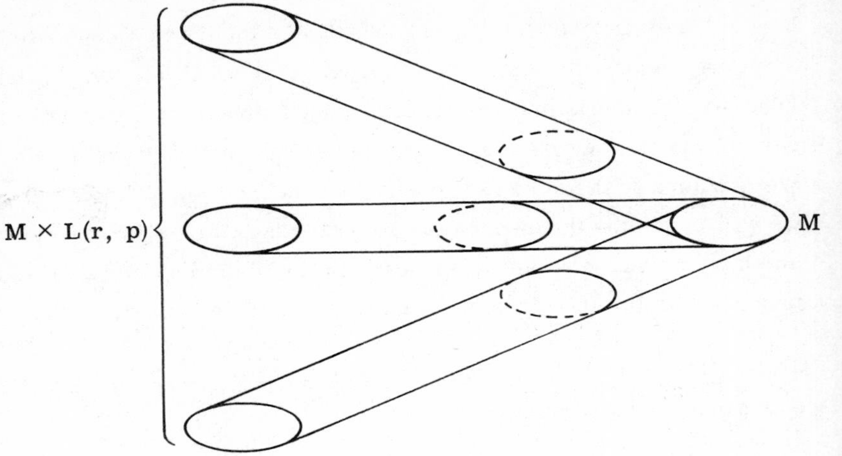

Fig. 6

Description of s

Let P be a ρ-manifold with singularity S(P). S(P) is an F_1-manifold, moreover the map

$$\Omega_{n-1} \otimes F_1 \xrightarrow{\otimes \phi_1} \Omega_{n-1} \otimes F_0,$$

can be described on generators as product with $L(r, \rho)$. Hence the image of S(P) in $\Omega_{n-1} \otimes F_0$ is represented by ∂U, where U is a regular neighbourhood of S(P) in P. This is bordant to zero in $\Omega_{n-1} \otimes F_0$ since it bounds cl (P - U). Thus the 'singularity' defines a map $s : \Omega_n(\rho) \to \mathrm{Tor}(\Omega_{n-1}, G)$ and it is surjective by reversing the above argument.

Exactness at $\Omega_n(\rho)$

Order 2 is trivial (an F_0-manifold has no singularity). To see exactness, suppose M is a ρ-manifold with S(M) bordant to zero as an F_1-manifold by a bordism W. Construct the bordism

$$W_1 = M \times I \cup W_r \times C(L(r, \rho))$$

where W_r is a typical component of W (labelled by r) and the union identifies the obvious subset of $M \times \{1\}$ with $\partial W_r \times C(L(r, \rho))$. W_1 is a ρ-bordism of M to an F_0-manifold. See Fig. 7.

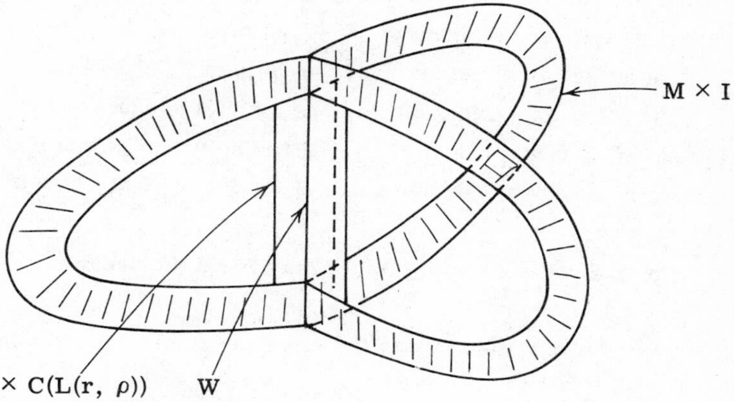

Fig. 7

2. COEFFICIENTS IN A LONGER RESOLUTION

In this section we will generalise the construction of §1 to give coefficients in resolutions of length ≤ 4.

Let G be an abelian group. A structured resolution ρ of G consists of

(a) a free resolution of G of length ≤ 4

$$0 \to F_3 \xrightarrow{\phi_3} F_2 \xrightarrow{\phi_2} F_1 \xrightarrow{\phi_1} F_0 \xrightarrow{\varepsilon} G \to 0$$

(b) a basis, B^p, for each F_p $(p = 0, 1, 2, 3)$

(c) for each $b^p \in B^p$ $(p = 1, 2, 3)$ we are given an unordered word, $w(b^p)$, representing the element $\phi_p(b^p)$. Precisely $w(b^p)$ is a finite set (of pairs)

$\{\delta_i b_i^{p-1} \mid i$ varies over a finite set $I(b^p)$, $\delta_i = +$ or $-$; $b_i^{p-1} \in B^{p-1}\}$

such that $\sum_i \delta_i b_i^{p-1} = \phi_p(b^p)$, where the sum denotes the element of F_{p-1} obtained from $w(b^p)$ by adding up all the coefficients belonging to the same element in B^{p-1}.

(d) For each $b^p \in B^p$ $(p = 2, 3)$ we are given a 'cancellation rule' defined as follows. Let $w(b^p) = \{\delta_i b_i^{p-1} : i \in I(b^p)\}$. Then order two of ρ implies that the formal sum $\sum_i \delta_i w(b_i^{p-1})$ is an unordered word representing the zero element of F_{p-2} in terms of the elements of B^{p-2}. The effect of δ_i is an inversion of sign iff $\delta_i =$ '-'.

The given cancellation rule consists of a procedure for pairing off the letters of $\sum_i \delta_i w(b_i^{p-1})$ in F_{p-2}. Precisely $c(b^p)$ is a partition of the letters of $\sum_i \delta_i w(b_i^{p-1})$ into pairs of the form $(\delta_j b^{p-2}, \delta_k b^{p-2})$ with $\delta_j \neq \delta_k$.

For the sake of simplicity we may write the set $I(b^p)$ in the form $\{1, \ldots, l : l = l(b^p)\}$.

If $0 \leq k \leq 3$ then ρ_k is the structured resolution of $\mathrm{Im}\, \phi_k$

$$\rho_k : 0 \to F_3 \xrightarrow{\phi_3} \cdots \to F_k \xrightarrow{\phi_k} \mathrm{Im}\, \phi_k \to 0$$

where the structure is that induced from ρ. Clearly $\rho_0 = \rho$.

For each quadruple (G, ρ, p, n), where G is an abelian group, ρ is a structured resolution of G; p, n are integers such that $-1 \leq p \leq 2$; $n \geq 0$, we shall construct

(a) a class of pl isomorphism types of polyhedra with extra structure, called the class of (ρ, p, n)-manifolds.

(b) a class $\mathcal{L}^p$ of $(\rho, p-1, p)$-manifolds for $p \geq 0$, called the class of (p, ρ)-links.

We start by defining a $(\rho, -1, n)$-manifold to be an F_0-manifold and in general (ρ, p, n)-manifolds will be defined from $(\rho, p-1, n)$-manifolds by 'killing' the class $\mathcal{L}^p$. The precise definition (an extension of that in §1) is given at the end of the construction of the classes $\mathcal{L}^p$.

The $(\rho, 2, n)$-manifolds will be called simply (ρ, n)-manifolds.

Construction of $\mathcal{L}^p$

For diagrams we refer to the examples given later. To construct $\mathcal{L}^0$ let $b^1 \in B^1$, $w(b^1) = \sum_{i=1}^{l} \delta_i b_i^0$. Give the set $\{b_1^0, \ldots, b_l^0\}$ the discrete topology and each point b_i $(i = 1, \ldots, l)$ the orientation δ_i. The resulting polyhedron will be called the 0-link associated to b^1 in ρ (or generated by b^1 in ρ) and will be written $L(b^1, \rho)$. We define the class $\mathcal{L}^0$ to consist of all polyhedra $L(b^1, \rho)$ with $b^1 \in B^1$, i.e. $\mathcal{L}^0 \equiv \{L(b^1, \rho) : b^1 \in B^1\}$. Now consider the join of $L(b^1, \rho)$ and the point b^1, written $b^1L(b^1, \rho)$, and give $b^1L(b^1, \rho) - b^1$ (= the open cylinder over $L(b^1, \rho)$) the orientation $- \!\!\rightarrow\!\! +$ (arrow departing from '-' sign). The join $b^1L(b^1, \rho)$, with the above orientation off b^1 and with the orientation '+' on b^1, will be referred to as the (oriented) cone over $L(b^1, \rho)$ with vertex b^1. In general it happens that different elements of B^1 may give rise to the same 0-link. Therefore, over the same link, there may be different cones, corresponding to different vertices generating the link.

To construct $\mathcal{L}^1$, let $b^2 \in B^2$; $w(b^2) = \sum_{i=1}^{l(b^2)} \delta_i b_i^1$. Construct the 0-link $L(b^2 : \rho_1)$ as in the previous case. Each b_i^1 generates a 0-link $L(b_i^1, \rho)$ $(i = 1, \ldots, l)$; consider the space $\bigcup_{i=1}^{l} \delta_i[b_i^1 L(b_i^1, \rho)] = L$ where $b_i^1 L(b_i^1, \rho)$ is the cone as defined above and δ_i changes all the orientations present in this cone iff $\delta_i =$ '-'. Let $\overline{w} = \sum \delta_i w(b_i^1) =$ $\{b_1^0, \ldots, b_t^0$ with the appropriate signs$\}$. Then $\overline{L} = \bigcup_{i=1}^{l} \delta_i L(b_i^1, \rho)$ is obtained by giving $\overline{w}$ the discrete topology and orientations according to the signs. Therefore the cancellation rule $c(b^2)$ gives a canonical way of joining the points of $\overline{L}$ in pairs by plugging in oriented 1-disks. Precisely suppose $\delta_j b_j^0$ is paired with $\delta_k b_k^0$. Then $\delta_j \neq \delta_k$ and we insert a 1-disk $[b_j^0, b_k^0]$ with orientation given according to the rule 'arrow departing from '+' sign'. Moreover we label the 1-disk by the unique element $b_j^0 = b_k^0 \in B^0$. The object which is obtained from L through the above identifications in $\overline{L}$ is called the 1-link associated to b^2 in ρ (or generated by b^2 in ρ) and it is written $L(b^2, \rho)$. The class of 1-links, $\mathcal{L}^1$, is defined by $\mathcal{L}^1 \equiv \{L(b^2, \rho) : b^2 \in B^2\}$.

It is clear that it is the given cancellation rule in ρ that makes the construction well defined. Different rules may give rise to completely different links.

We think of $L(b^2, \rho)$ as a one-dimensional stratified set in which the intrinsic j-stratum $(j = 0, 1)$ consists of a disjoint union of j-disks, each disk is oriented and labelled by one element of B^{1-j}; the 0-stratum is the 0-link generated by b^2 in ρ_1.

It is clear how to define the cone $b^2L(b^2, \rho)$: topologically $b^2L(b^2, \rho)$ is the usual cone over $L(b^2, \rho)$ with vertex b^2, the subcone over the 0-stratum of $L(b^2, \rho)$ is given an orientation outside b^2 as in the previous step; the subcone over the 1-stratum has the orientation given by the cartesian product:

(1-stratum) $\times [-; +)$ where $[-; +)$ is the half open 1-disk oriented 'from - to +'. Finally the vertex b^2 has the orientation +. Each stratum of $b^2L(b^2, \rho)$ is labelled in the obvious way. Now, as before, it may well happen that different elements in B^2 generate the same 1-link and therefore over the same link there may be different cones.

Now suppose W is an oriented manifold, then we can form the topological product $W \times b^2L(b^2, \rho)$. From now on we think of the above product as having the following additional structure: three intrinsic strata, namely $W \times b^2$, $W \times L_0$, $W \times L_1$, L_j being the intrinsic j-dimensional stratum of $L(b^2, \rho)$ $(j = 0, 1)$; a labelling on each stratum obtained from the labelling of the second factor; the product orientation on each stratum.

We are now left with the case $p = 2$. Let $b^3 \in B^3$. Consider the 1-link associated to b^3 in ρ_1 and construct a trivial normal bundle system with base $L(b^3, \rho_1)$ as follows. If $\delta_i b_i^2$ is a vertex of $L(b^3, \rho_1)$ then put $\delta_i L(b_i^2, \rho)$ as the fibre at that vertex. The part of $L(b^3, \rho_1)$ which remains unclothed consists of a disjoint union of closed 1-disks. Let D, labelled by $b^1 \in B^1$, be one of such disks. The restriction of the normal bundle to ∂D is $\partial D \times L(b^1, \rho)$; therefore we can extend the bundle by plugging in $D \times L(b^1, \rho)$. As a result of clothing the 1-stratum of $L(b^3, \rho_1)$ we are left with a polyhedron, whose boundary consists of 1-spheres, labelled by elements of B^0. Then plug in an oriented labelled 2-disk for each sphere and get the required link $L(b^3, \rho)$.

The cone $b^3L(b^3, \rho)$ and the product $W \times b^3L(b^3, \rho)$, where

W is an oriented manifold, are defined as in the previous cases.

We now define (ρ, p, n)-manifolds (without boundary) inductively on p as follows. A (ρ, p, n)-manifold is a polyhedral pair $M \supset S(M)$ with labellings and extra structure such that

(a) $M - S(M)$ is a $(\rho, p-1, n)$-manifold

(b) $S(M)$ is an F_{p+1}-manifold

(c) (Trivialised stratification condition) For each component V of $S(M)$ labelled by $b^{p+1} \in B^{p+1}$, there is given a regular neighbourhood N of V in M and an isomorphism

$$h : (N, V) \to (V \times {}^{b^{p+1}}L(b^{p+1}, \rho), V \times b^{p+1})$$

(d) h is an isomorphism of $(\rho, p-1, n)$-manifolds off V.

Remarks 2.1. 1. It is obvious how to give a p-link the required extra structure to make it into a $(\rho, p-1, p)$-manifold (and this completes all the definitions).

2. In all the above definitions we have only used the fact that the resolution ρ has order two.

3. The definition extends to yield (ρ, n)-manifolds with boundary in the obvious way.

Examples 2.2. 1. A short resolution gives rise to an obvious structured resolution (the cancellation rule gives no information in this case). Then the notions of ρ-manifold defined in §1 and §2 coincide.

2. $\rho : 0 \to \operatorname{Ker} \phi_2 \xrightarrow{\phi_3} F \operatorname{Ker} \phi_1 \xrightarrow{\phi_2} F \operatorname{Ker} \varepsilon \xrightarrow{\phi_1} FZ_3 \to Z_3 \to 0$

Let $b^3 \in \operatorname{Ker} \phi_2$ be such that

$$w(b^3) = -b^2_1 + b^2_2 + b^2_3$$

$$w(b^2_1) = -b^1_{11} + b^1_{12} \in F \operatorname{Ker} \varepsilon$$

$$w(b^2_2) = b^1_{21} - b^1_{22} + b^1_{23} \in F \operatorname{Ker} \varepsilon$$

$$w(b^2_3) = -b^1_{31} + b^1_{32} + b^1_{33} \in F \operatorname{Ker} \varepsilon$$

$$w(b^1_{11}) = b^0_1 + b^0_2$$

$$w(b^1_{12}) = -b^0_1 - b^0_2$$

$$w(b^1_{21}) = b^0_1 + b^0_2$$

$$w(b^1_{22}) = b^0_0 + b^0_{11} + b^0_{12} + b^0_{21} + b^0_{22}$$

$$w(b^1_{23}) = b^0_0 + b^0_1 + b^0_2$$

$$w(b^1_{31}) = b^0_0 + b^0_1 + b^0_2$$

$$w(b^1_{32}) = b^0_0 + b^0_{11} + b^0_{12} + b^0_{21} + b^0_{22}$$

$$w(b^1_{33}) = -b^0_1 - b^0_2$$

$b^0_{11} = b^0_{12} = b^0_1 \in \mathbf{FZ}_3$; $b^0_{21} = b^0_{22} = b^0_2 \in \mathbf{FZ}_3$; $\varepsilon(b^0_0) = 0 \in \mathbf{Z}_3$; $\varepsilon(b^0_1) = 1 \in \mathbf{Z}_3$; $\varepsilon(b^0_2) = 2 \in \mathbf{Z}_3$

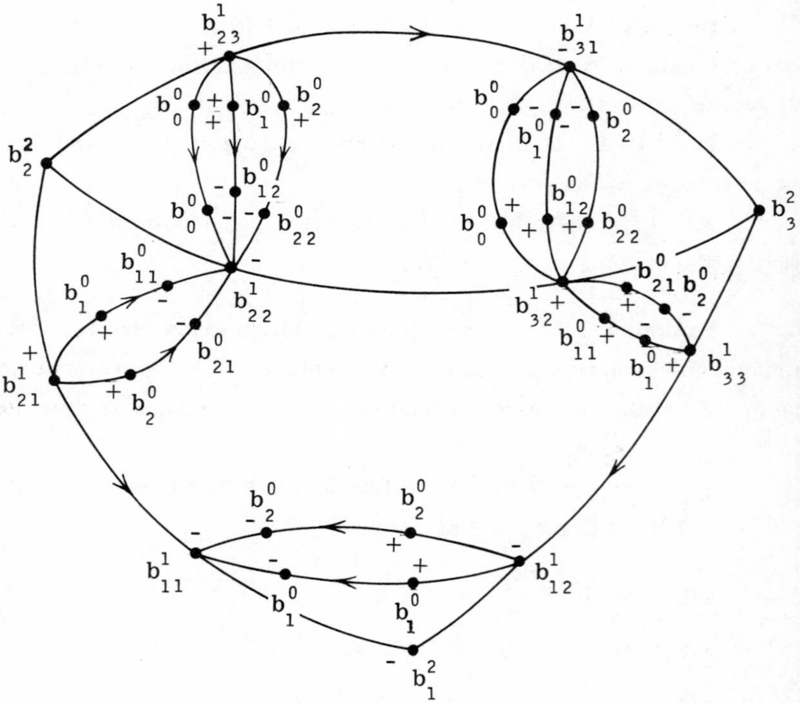

Fig. 8

Figure 8 shows the construction of $L(b^3, \rho)$, the cancellation rules being suggested by the diagram itself.

3. $\rho : 0 \to \text{Ker}\ \phi_2 \to \text{F Ker}\ \phi_1 \xrightarrow{\phi_2} \text{F Ker}\ \varepsilon \xrightarrow{\phi_1} FZ_5 \to Z_5 \to 0.$
Let b^2 be a basis element of F Ker ϕ_1. Suppose

$$w(b^2) = b^1_1 + b^1_2 + b^1_3 + b^1_4$$
$$w(b^1_1) = b^0_{11} + b^0_{12} + b^0_{13}$$
$$w(b^1_2) = -b^0_{21} + b^0_{22} + b^0_{23}$$
$$w(b^1_3) = -b^0_{31} - b^0_{32} + b^0_{33}$$
$$w(b^1_4) = -b^0_{41} - b^0_{42} - b^0_{43}$$
$$\varepsilon(b^0_{11}) = \varepsilon(b^0_{43}) = 2;\ \varepsilon(b^0_{12}) = \varepsilon(b^0_{42}) = 3;$$
$$\varepsilon(b^0_{13}) = \varepsilon(b^0_{21}) = \varepsilon(b^0_{33}) = \varepsilon(b^0_{41}) = 0;$$
$$\varepsilon(b^0_{22}) = \varepsilon(b^0_{32}) = 1;\ \varepsilon(b^0_{23}) = \varepsilon(b^0_{31}) = 4.$$

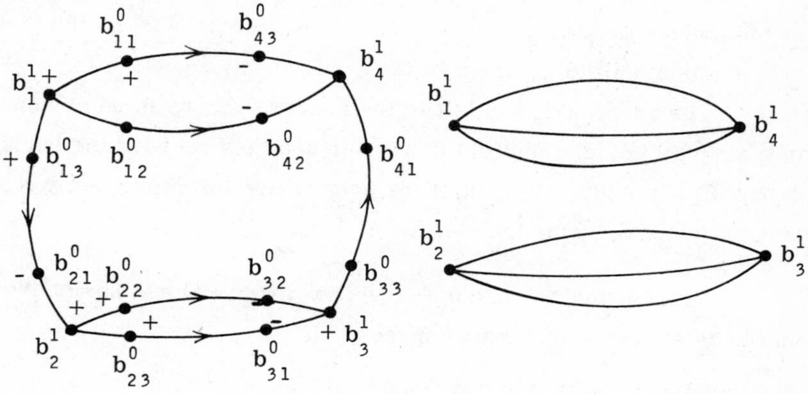

Fig. 9

Figure 9 shows two possible links associated to b^2 in ρ corresponding to different cancellation rules. This completes the examples.

Now let M be a (ρ, n)-manifold and M' a (ρ, n')-manifold. An <u>embedding</u> $f : M' \to M$ is a locally flat stratified embedding between the underlying polyhedra, which is compatible with the labelling and the trivialisations. If $n' = n$, then f may be orientation preserving or orientation-reversing. In the following, unless otherwise stated, a codimension zero embedding will always be assumed to be orientation preserving. A submanifold of a (ρ, n)-manifold M is a subset $M_0 \subset M$ together with an embedding $f : M' \hookrightarrow M$ (of ρ-manifolds) such that $f(M') = M_0$. If M is a (ρ, n)-manifold, $-M$ denotes the (ρ, n)-manifold obtained from M by reversing all the orientations; (ρ, n)-manifolds have the following properties.

Proposition 2. 3. 1. If M is a (ρ, n)-manifold, $M \times I$ has a natural structure of $(\rho, n+1)$-manifold, obtained by crossing the structure of M with I, it is clear that $\partial(M \times I) \cong M \cup -M \cup \partial M \times I$.

2. If M, M' are (ρ, n)-manifolds and M_0, M'_0 are $(\rho, n-1)$-submanifolds of $\partial M, \partial M'$ respectively such that $M_0 \overset{g}{\cong} -M'_0$, then $M \cup_g M'$ is a (ρ, n)-manifold with boundary isomorphic to $Cl(\partial M \setminus M_0 \cup_g \partial M' \setminus M'_0)$.

3. Let M be a (ρ, n)-manifold and $X \subset M$. Let N be a regular neighbourhood of X in M. Then N can be given the structure of a (ρ, n)-manifold in an essentially unique way.

The proof of 2. 3 is left to the reader. There is an obvious notion of a singular (ρ, n)-manifold in a space and thus we have the bordism group $\Omega_n(X, A; \rho)$. The following proposition follows directly from proposition 2. 3, using the proof of II 3. 1.

Proposition 2. 4. $\{\Omega_n(\; , \; ; \rho)\}$ is a generalised homology theory on the category of topological spaces.

The universal coefficient sequence

Proposition 2. 5. For each integer $n \geq 0$ and each pair (X, A) there exists a short exact sequence

$$0 \to H_0(\rho, \Omega_n(X; A)) \xrightarrow{l} \Omega_n(X; A; \rho) \xrightarrow{s} H_1(\rho, \Omega_{n-1}(X, A)) \to 0$$

which is natural in (X, A).

Proof. First of all we anticipate that the proof consists of geometrical arguments involving only (ρ, n)-manifolds and their stratifications. In the constructions, the maps into (X, A) do not play an essential role and so, for the sake of simplicity, we shall assume $(X, A) = (\text{point}, \emptyset)$.

Description of the map s

It involves a resolution of singularities. Let M be a closed (ρ, n)-manifold. We are going to show that there exists a (ρ, n)-manifold $\tilde{M}$, bordant to M and having no singularities in codim > 1. Let SM be the last stratum of M. Then, by definition of (ρ, n)-manifold, SM is an F_p-manifold, where SM has codimension p. We need the following:

Lemma 2.6. Suppose that $[SM] = [S]$ as F_p-manifolds. Then M is bordant to a (ρ, n)-manifold Q, whose last stratum is S, by a bordism R whose last stratum is still in codimension p.

Proof. Consider the following spaces:

$M \times I'$, where $I' = [0, -1]$;

SR = any bordism between -SM and -S:

assume that SR consists of a set of equally labelled components, with label, say, $b^p \in B^p$;

$\nu(-SM)$ = normal bundle of -SM in $-M = M \times \{-1\}$;

$SR \times L(b^p, \rho)$.

We have: $\nu(-SM) \subset M \times \{-1\}$; $\nu(-SM) \subset SR \times L(b^p, \rho)$. So we can form the identification space:

$$R = SR \times L(b^p, \rho) \underset{\nu(-S(M))}{\frown} M \times I'$$

which provides the required bordism.

If SR has many labelling elements, we perform the above construction simultaneously on every set of equally labelled components. The last stratum of R is SR and has codimension p. See Fig. 10.

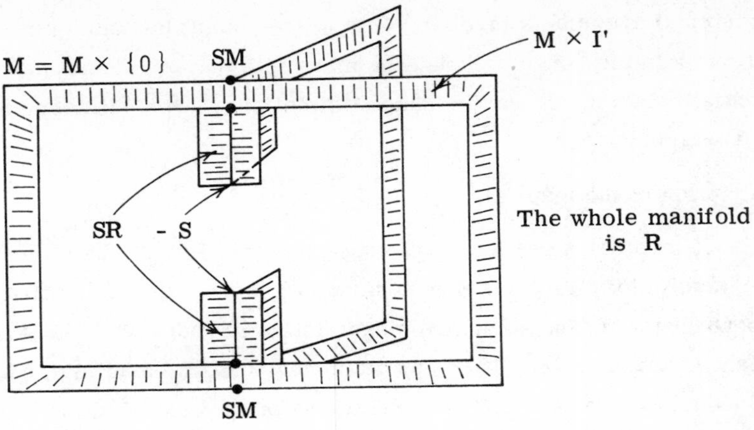

Fig. 10

Remark 2. 7. If we can choose $S = \emptyset$ then the above construction gives a resolution of the low dimensional singularities of M. In other words, when the low dimensional singularities of a (ρ, n)-manifold M bound (in a labelled sense), they can be resolved by means of a bordism having the same kind of singularities as M. (Cf. proof of exactness in 1. 2.)

Proof of Proposition 2. 5 (continued). Now let us look at the image of [SM] through the morphism

$$\tilde{\phi}_p : \Omega_{n-p} \otimes F_p \xrightarrow{\text{id} \otimes \phi_p} \Omega_{n-p} \otimes F_{p-1}.$$

We have, for each component $V_i \otimes b_i^p \subset SM$, $\tilde{\phi}_p([V_i] \otimes b_i) = [V_i] \otimes w(b_i^p)$ $[V_i] \otimes \sum_j \delta_j b_j^{p-1} = \sum_j [\delta_j V_i] \otimes b_j^{p-1}$: this is nothing else than the bordism class of the boundary of the complement of a regular neighbourhood of V_i in the (n-p+1)-stratum of M. Therefore the image of [SM] is the bordism class of the boundary of the complement of a regular neighbourhood SM in the (n-p+1)-stratum of M and, as such, is the zero element of $\Omega_{n-p} \otimes F_{p-1}$. Now, because $p > 1$, the sequence:

$$\Omega_{n-p} \otimes F_{p+1} \xrightarrow{\tilde{\phi}_{p+1}} \Omega_{n-p} \otimes F_p \xrightarrow{\phi_p} \Omega_{n-p} \otimes F_{p-1}$$

is exact and so there exists an element $[SW] \in \Omega_{n-p} \otimes F_{p+1}$ such that $\tilde{\phi}_{p+1}[SW] = [SM]$. Suppose first that SW is a set of components all labelled by $b^{p+1} \in B^{p+1}$. We can always reduce to the case $SM = SW \otimes w(b^{p+1})$, because if SM is only bordant to $SW \otimes w(b^{p+1})$: $SM \sim SW \otimes w(b^{p+1})$, then, by Lemma 2.6, M can be replaced by another (ρ, n)-manifold $\overline{M}$ such that:

(a) $\overline{M}$ is bordant to M by means of a bordism with singularities up to codimension p only

(b) $S\overline{M} = SW \otimes w(b^{p+1})$.

Therefore assume $SM = SW \otimes w(b^{p+1})$. If $SW \sim \emptyset$ we are reduced to the case of Remark 3 and we know how to solve the singularities then. So assume $SW \neq \emptyset$ and take the following spaces:

$M \times I$, where $I = [0, -1]$

$-(SW \times b^{p+1}L(b^{p+1} : \rho))$

$\nu(-SM)$ = normal bundle system of $-SM$ in $M \times \{-1\} = -M$.

Then we have $\nu(-SM) \subset M \times I$ and $\nu(-SM) \subset -(SW \times b^{p+1}L(b^{p+1}, \rho)$.
The identification space

$$W = -(SW \times b^{p+1}L(b^{p+1}, \rho)) \underset{\nu(-S(M))}{\smile} M \times I$$

realises a bordism between $M = M \times \{0\}$ and a (ρ, n)-manifold M' whose last stratum has dimension $n - p + 1$. The singularities SM have been resolved up to bordism. In general, if SW is of the form $SW = \sum_k (SW)_k \otimes b_k^{p+1}$, then one performs the above construction simultaneously on all terms $(SW)_k \otimes b_k^{p+1}$ and gets the desired manifold M'. See Fig. 11.

We remark that in order to get rid of singularities in codimension p we have used bordisms, which have singularities up to codimension p + 1.

So now we have a well defined procedure to solve the singularities of a (ρ, n)-manifold M, stratum by stratum, starting from the last one and going up by one dimension each time, until we are left with a (ρ, n)-manifold, $\tilde{M}$, which is bordant to M and has singularities $S\tilde{M}$ in codimension one at most. In general we cannot solve $S\tilde{M}$ as above, because

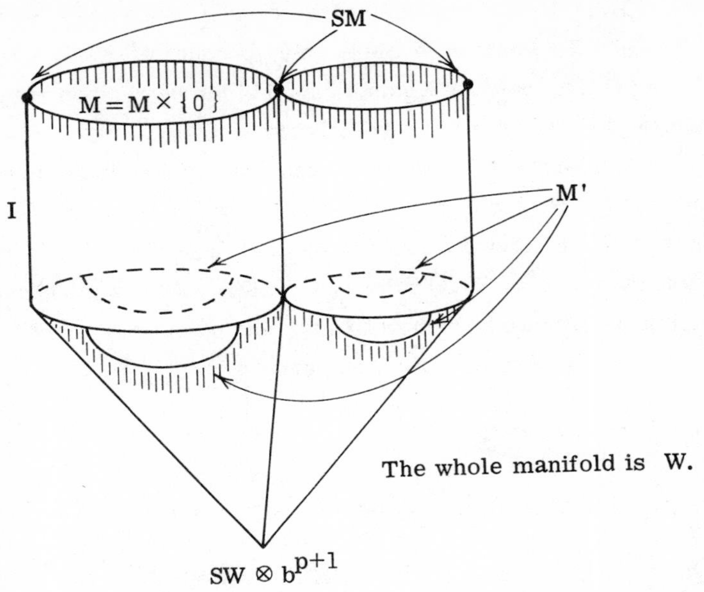

Fig. 11

the sequence

$$\Omega_{n-1} \otimes F_2 \xrightarrow{\tilde{\phi}_2} \Omega_{n-1} \otimes F_1 \xrightarrow{\tilde{\phi}_1} \Omega_{n-1} \otimes F_0$$

is not necessarily exact. However $\tilde{\phi}_1[S\tilde{M}] = 0$ in $\Omega_{n-1} \otimes F_0$ because it is the bordism class of the boundary of the complement of a regular neighbourhood of $S\tilde{M}$ in $\tilde{M}$.

The next lemma is important in what follows.

Lemma 2.8. Suppose that M is a $(\rho, n+1)$-manifold with boundary ∂M. Then, if ∂M has singularities up to codimension p and M has singularities up to codimension $p + h$, there is a (ρ, n)-bordism with boundary, W, ∂W, between M, ∂M and N, ∂N such that

(a) W has singularities up to codimension $p + h + 1$ at most,

(b) ∂W has singularities up to codimension p,

(c) N has singularities up to codimension $p + 1$.

Proof. The intrinsic stratum of M in codimension $p + 1$ is a closed polyhedron, because the singularities of ∂M are in codimension p at most. Therefore the above construction for solving the singularities can be applied, essentially unaltered, to solve the codimension $p + 2$ stratum of M, ∂M by a bordism with boundary, W, ∂W. No new singularities are created along the boundary during this process.

Proof of Proposition 2.5 (continued). Each $M \in [M]_\rho$ determines an element of Ker ϕ_1, namely $[S\tilde{M}]$. This assignment depends on the representative M; in fact, if $M_1 \sim M$, then $S\tilde{M}_1$ may be bordant to $S\tilde{M}$ by means of a bordism, V, with singularities and therefore in general $[S\tilde{M}_1] \neq [S\tilde{M}]$. However, by Lemma 2.8, the singularities of V can be assumed to have codimension one at most and we can certainly write: $[S\tilde{M}] = [S\tilde{M}_1] + [N]$ where $[N] \in \mathrm{Im}\,\tilde{\phi}_2$. Thus there is a well defined map:

$$\begin{aligned} s : \Omega_n(\rho) &\to H_1(\rho, \Omega_{n-1}) \\ [M] &\mapsto [S\tilde{M}] + \mathrm{im}\,\tilde{\phi}_2 . \end{aligned}$$

It is straightforward to check that s is a morphism of groups.

s is an <u>epimorphism.</u> Take $[V] + \mathrm{im}\,\tilde{\phi}_2 \in H_1(\rho, \Omega_{n-1})$; $\tilde{\phi}_1[V] = 0$. Suppose V constantly labelled by $b^1 \in B^1$; then $V \otimes w(b^1)$ bounds in $\Omega_{n-1} \otimes F_0$, i.e. $V \otimes w(b_1) = \partial\tilde{V}$. Take a copy of V and label it by b^1; $\partial\tilde{V}$ consists of a number of copies of V (non constantly labelled in general); identify each copy with $V \otimes b^1$. $\tilde{V}$, with the above identification on its boundary, becomes a (ρ, n)-manifold $\tilde{W}$ with singularities in codimension one only. Therefore, to each manifold V, representing an element in $H_1(\rho, \Omega_{n-1})$ we are able to associate a (ρ, n)-manifold $\tilde{W}$, representing an element in $\Omega_n(\rho)$, such that $s[\tilde{W}] = [V] + \mathrm{im}\,\tilde{\phi}_2$. In fact, if $\tilde{W}' \in [\tilde{W}]$ we can assume, by Lemma 2.8, that $\tilde{W}'$ has singularities $S\tilde{W}'$ in codimension at most one and that there exists a bordism $N : \tilde{W}' \sim \tilde{W}$ with singularities SN in codimension two at most. Then $[S\tilde{W}] - [S\tilde{W}'] = \tilde{\phi}_2[SW]$ and so s is an epimorphism.

Description of the map l

Define a map $\tilde{l} : \Omega_n \otimes F_0 \to \Omega_n(\rho)$ by $\tilde{l}[M] = [M]$; $\tilde{l}$ is a well defined homomorphism; so we have the sequence

$$\Omega_n \otimes F_1 \xrightarrow{\tilde{\phi}_1} \Omega_n \otimes F_0 \xrightarrow{\tilde{l}} \Omega_n(\rho).$$

$\tilde{l}\tilde{\phi}_1 = 0$ because let $[W] = \tilde{l}\tilde{\phi}_1[M]$ and suppose M constantly labelled by $b^1 \in B^1$. Then take $M \times b^1L(b^1, \rho)$ and observe that $\partial(M \times b^1L(b^1, \rho))$ is bordant to W. So stick the two bordisms together and get a bordism of W to the empty set by means of a $(\rho, n+1)$-manifold with codimension one singularities.

Assume now $\tilde{l}([M]) = 0$. Pick a representative V of $\tilde{l}([M])$: there exists a bordism, N, of V to ϕ such that N has singularities SN in codimension one at most. We claim that $\tilde{\phi}_1[SN] = [M]$. In fact remove from N a regular neighbourhood of SN in N to get the required bordism between M and $\tilde{\phi}_1(SN)$. Thus we have proved that the sequence above is exact; which is enough to ensure the existence of a monomorphism $l : H_0(\rho, \Omega_n) \to \Omega_n(\rho)$ induced by $\tilde{l}$.

Now it only remains to prove exactness at $\Omega_n(\rho)$.

$sl = 0$: $s\tilde{l}[M] = 0$, because $[M]$ has no singularities; hence $sl = 0$.

$\text{Ker } s \subset \text{im } l$: let $[M] \in \Omega_n(\rho)$ and assume, without loss of generality, that M has codimension one singularities SM; $s[M]_\rho = 0$ means that $[SM] \in \text{im } \tilde{\phi}_2$. But then SM can be re-solved up to bordism; therefore $[M] = [M']$ where M' is without singularities and hence determines an element of $\Omega_n \otimes F_0$ whose image through $\tilde{l}$ is $[M]$. Thus $\text{Ker } s \subset \text{im } \tilde{l} = \text{im } l$.

The proof of the proposition is now complete.

Remark 2.9. We have seen how the exactness of ρ is used in the proof of the universal-coefficient theorem.

As pointed out before, if ρ is any based ordered chain complex augmented over G, then the theory $\Omega_*(-, \rho)$ can be defined in the same way. But now the singularities in codimension greater than one are not necessarily solvable; they give rise to the E^2-term of a spectral sequence running:

$$H_p(\rho, \Omega_q(-)) \Rightarrow \Omega_*(-; \rho).$$

This spectral sequence collapses to the universal coefficient formula when ρ is exact.

3. FUNCTORIALITY

The classes of links constructed in Section 2 summarize the whole structure of the resolution ρ geometrically. Therefore, from now on, we refer to ρ as a linked resolution.

If ρ, ρ' are linked resolutions of G, G' respectively, a chain map $f : \rho \to \rho'$ is said to be a map of linked resolutions (or simply linked map) if f is based and link preserving, i. e. :-

(a) $f(b^p) \in B'^p$ for each $b^p \in B^p$.

(b) Let $b^p \in B^p$. If we relabel each stratum of the link $L(b^p, \rho)$ according to f and if $f(L(b^p, \rho))$ denotes the resulting object, then $f(L(b^p, \rho)) = L(fb^p, \rho')$.

So there is a category, $\mathcal{C}$, whose objects are linked resolutions $\rho \xrightarrow{\varepsilon} G$ and whose morphisms are linked maps. If $\mathcal{A}b_*$ is the category of graded abelian groups, we have the following

Proposition 3.1. $\Omega_*(X, A; \rho)$ is a functor $\mathcal{C} \to \mathcal{A}b_*$. (For the sake of simplicity we disregard the topological component of $\Omega_*(-; -)$.)

Proof. Let $\tau : \rho \to \rho'$ be a morphism of $\mathcal{C}$. If $[M]_\rho \in \Omega_n(-; \rho)$, we associate a (ρ', n)-manifold, $\tau(M)$, to M by relabelling all the strata of M according to the based map τ. The correspondence $[M]_\rho \to [\tau(M)]_\rho$ gives a well defined natural transformation of theories $\tau_* : \Omega_*(-; \rho) \to \Omega_*(-; \rho')$ and the functorial properties are clear.

Corollary 3.2. If the linked map $\tau : \rho \to \rho'$ is a homotopy equivalence, then τ_* is an isomorphism.

Proof. This is an easy consequence of the universal-coefficient theorem. There is a commutative diagram

$$\begin{array}{ccccccccc}
0 \to & H_0(\rho, \Omega_n(X, A)) & \to & \Omega_n(X, A; \rho) & \to & H_1(\rho, \Omega_{n-1}(X, A)) & \to 0 \\
& \Big\downarrow {\scriptstyle H_0(\tau, \Omega_n(X, A))} & & \Big\downarrow {\scriptstyle \tau_*} & & \Big\downarrow {\scriptstyle H_1(\tau, \Omega_{n-1}(X, A))} & \\
0 \to & H_0(\rho', \Omega_n(X, A)) & \to & \Omega_n(X, A; \rho) & \to & H_1(\rho, \Omega_{n-1}(X, A)) & \to 0
\end{array}$$

in which the side-morphisms are isomorphisms, because τ is a homotopy equivalence. Therefore τ_* is also an isomorphism.

If $G \in \mathcal{A}b$, <u>a truncated linked (t. l.) resolution of</u> G is an exact sequence

$$F_2 \xrightarrow{\phi_2} F_1 \xrightarrow{\phi_1} F_0 \xrightarrow{\varepsilon} G \to 0$$

satisfying conditions (a), (b), (c), (d) in the definitions of structured resolution given in Section 2. A linked map between t. l. resolutions is defined as in the non-truncated case and there is a category, $\tilde{\mathcal{C}}$, whose objects are t. l. resolutions and whose morphisms are linked maps.

In the following, for each $\rho \in \tilde{\mathcal{C}}$ and each topological pair (X, A), we construct a graded abelian group $\{\tilde{\Omega}_*(X, A; \rho)\}$ which is a functor on $\mathcal{P} \times \mathcal{A}b$ ($\mathcal{P}$ = category of topological pairs). Fix a $\rho' \in \mathcal{C}$ obtained from ρ by choosing a based kernel of ϕ_2. A singular (ρ, n)-cycle in (X, A) is a pair (M, f) consisting of a (ρ', n)-manifold M and a map $f : (M, \partial M) \to (X, A)$ such that M has at most two intrinsic strata labelled by elements of B^0 and B^1. A <u>singular</u> (ρ, n)-<u>cycle</u> (M, f) is said to <u>bord</u> if there exists a $(\rho', n+1)$-manifold W and a map $F : W \to X$ for which

(a) M is a submanifold of ∂W

(b) $F|M = f$ and $F(\partial W \setminus M) \subset A$

(c) M has at most three intrinsic strata (labelled by elements of B^0, B^1, B^2).

W is called a <u>bordism.</u> Define $-(M, f) = (-M, f)$. Two singular (ρ, n)-cycles (M_1, f_1), (M_2, f_2) are <u>bordant</u> if the disjoint union $(M_1 \cup -M_2, f_1 \cup f_2)$ bords in (X, A). Bordism is an equivalence relation in the set of singular (ρ, n)-cycles of (X, A). Denote the bordism class of (M, f) by $[M, f]$ and the set of all such bordism classes by $\tilde{\Omega}_n(X, A; \rho)$. An abelian-group structure is given in $\tilde{\Omega}_n(X, A; \rho)$ by disjoint union.

Obviously the above definition of $\tilde{\Omega}_n(X, A; \rho)$ is independent of the chosen $\rho' \in \mathcal{C}$.

Now we fix our attention on a particular t. l. resolution, called the *canonical resolution* of G and written γ. It is defined as follows:

$$\begin{array}{ccccccccc}
\gamma : \Gamma_2 & \xrightarrow{\ \phi_2\ } & & \Gamma_1 & \xrightarrow{\ \phi_1\ } & & \Gamma_0 & \xrightarrow{\ \varepsilon\ } G \longrightarrow 0 \\
& \searrow^{\psi_2} & \nearrow_{\partial_2} & & \searrow^{\psi_1} & \nearrow_{\partial_1} & & \\
& & F\,\mathrm{Ker}\,\phi_1 & & & F\,\mathrm{Ker}\,\varepsilon & &
\end{array}$$

where

Γ_0, F Ker ε, F Ker ϕ_1 are the free abelian groups on G, Ker ε, Ker ϕ_1 respectively

∂_1 is the obvious map; $\Gamma_1 = FB^1$

$B^1 = \{(f, w(f)) \mid f \in \mathrm{Ker}\,\varepsilon \subset F\,\mathrm{Ker}\,\varepsilon$ and $w(f)$ is a word expressing $\partial_1 f$ in terms of the elements of $G \subset \Gamma_0\}$;

$\psi_1(f, w(f)) = f$ and $\phi_1 = \partial_1 \psi_1$

$\Gamma_2 = FB^2$; $B^2 = \{h; w(h), c(h) \mid h \in \mathrm{Ker}\,\phi_1$, $w(h)$ is a word expressing $\partial_2(h)$, $c(h)$ is a cancellation rule associated to $h\}$. ψ_2 and ϕ_2 are defined similarly

γ has canonical bases G, B^1, B^2 and a canonical structure in which $(h, w(h), c(h))$ has $(w(h), c(h))$ as its structure.

Lemma 3.3. *If $\phi : G \to G'$ is a homomorphism of abelian groups, $\rho \xrightarrow{\varepsilon} G$ is a t. l. resolution of G, γ' is the canonical resolution of G'; then ϕ extends to a linked map $\tilde{\phi} : \rho \to \gamma'$ in a canonical way.*

Proof. Let $\rho = \{F_p, \phi_p\}$, $\gamma' = \{\Gamma'_p, \phi'_p\}$. We proceed by induction on p. Write $(\tilde{\phi}) = (\tilde{\phi}_0, \tilde{\phi}_1, \tilde{\phi}_2)$. For $p = 0$, put $\tilde{\phi}_0(b^0) = \phi\varepsilon(b^0)$ for each $b^0 \in B^0$. Inductively, let $b^p \in B^p$. Then $\tilde{\phi}_{p-1}\phi_p(b^p) \in \mathrm{Ker}\,\phi'_{p-1}$ and therefore it determines a basis element, b'^p, in F Ker ϕ'_{p-1}; b'^p has a canonical word $w(b'^p)$ and cancellation rule $c(b'^p)$ induced from those of b^p through the map $(\tilde{\phi}_{p-1}, \tilde{\phi}_{p-2})$. Therefore the assignment $b^p \to (b'^p, w(b'^p), c(b'^p))$ defines $\tilde{\phi}_p$ with the required properties.

Lemma 3.4. $\tilde{\Omega}_*(X, A; \rho)$ gives a functor $\mathcal{P} \times \tilde{\mathcal{C}} \to \mathcal{A}b_*$.

Proof. Functoriality on $\mathcal{P}$ is obvious.

If $\tau : \rho \to \rho'$ is a morphism of $\tilde{\mathcal{C}}$ and $[M]_\rho \in \tilde{\Omega}_n(-; \rho)$, let $\tau(M)$ be the (ρ', n)-manifold associated to M by relabelling each stratum according to τ. The correspondence $[M]_\rho \to [\tau(M)]_\rho$ gives the required natural transformation $\tau_* : \tilde{\Omega}_*(-; \rho) \to \tilde{\Omega}_*(-; \rho')$.

Corollary 3.5. $\tilde{\Omega}_*(X, A; \gamma)$ gives a functor $\mathcal{P} \times \mathcal{A}b \to \mathcal{A}b_*$.

Proof. Functoriality on $\mathcal{P}$ is obvious.

To each $G \in \mathcal{A}b$ assign $\tilde{\Omega}_*(X, A; \gamma)$ where γ is the canonical resolution of G; to each morphism $\phi : G \to G'$, $\phi \in \mathcal{A}b$, assign the homomorphism $\tilde{\phi}_* : \tilde{\Omega}_*(X, A; \gamma) \to \tilde{\Omega}_*(X, A; \gamma')$, where $\tilde{\phi}$ is the canonical extension of ϕ described in Lemma 3.3 and $\tilde{\phi}_*$ is the induced homomorphism described in Lemma 3.4.

In view of the previous corollary we shall write $\tilde{\Omega}_*(X, A; G)$ instead of $\tilde{\Omega}(X, A; \gamma)$ and ϕ_* instead of $\tilde{\phi}_*$. A (γ, n)-cycle [bordism] will also be called a (G, n)-manifold [(G, n)-bordism].

A linked resolution of an abelian group G

$$\rho : 0 \to F_3 \xrightarrow{\phi_3} F_2 \xrightarrow{\phi_2} F_1 \xrightarrow{\phi_1} F_0 \xrightarrow{\varepsilon} G \to 0$$

is said to be p-canonical $(1 \le p \le 3)$ if

(a) $F_{p-1} \to \dots \xrightarrow{\varepsilon} G \to 0$ is the canonical resolution of G of length $p - 1$, i.e. $F_i = \Gamma_i$, $0 \le i \le p - 1$, and the morphisms ϕ_i are the same as in the definition of γ.

(b) $F_{p+1} = F_{p+2} = 0$.

Let $G \in \mathcal{A}b$ and $\rho_G \in \mathcal{C}$ a short linked resolution of G, (i.e. $F_2 = F_3 = 0$). Then we have the homology theory $\{\Omega_*(X, A; \rho_G), \partial\}$ defined in Section 1. We also have the graded functor $\tilde{\Omega}_*(X, A; G)$: $\mathcal{P} \to \mathcal{A}b_*$ constructed at the beginning of this section. It follows from Lemma 3.3 that there is a canonical extension of the $id : G \to G$ to a linked map $\tilde{id} : \rho_G \to \gamma$. The latter induces a natural transformation of functors $t_G : \Omega_*(X, A; \rho_G) \to \tilde{\Omega}_*(X, A; G)$ obtained by relabelling the ρ_G-manifolds according to $\tilde{id}$. The next theorem is the main step towards functoriality.

Theorem 3.6. $t_G : \Omega_*(X, A; \rho_G) \to \tilde{\Omega}_*(X, A; G)$ *is a natural equivalence of functors.*

Proof. Let ρ_i be an i-canonical resolution for $i = 2, 3$. Then there is a commutative diagram

$$\begin{array}{ccccc} \tilde{\Omega}_*(X, A; G) & \xleftarrow{t_G} & \Omega_*(X, A; \rho_G) & \xrightarrow{t_{1,2}} & \Omega_*(X, A; \rho_2) \\ & \searrow^{\alpha} & \downarrow{\scriptstyle t_{1,3}} & \swarrow_{t_{2,3}} & \\ & & \Omega_*(X, A; \rho_3) & & \end{array}$$

where $t_{i,j}$ and α are the natural transformations obtained in the usual way by relabelling the cycles according to the canonical liftings of $id : G \to G$. By Corollary 3.2, $t_{i,j}$ is an isomorphism $(1 \leq i < j \leq 3)$. Therefore, in order to prove the theorem, we only need to show that α is an isomorphism.

α is an epimorphism: it follows from commutativity and the fact that $t_{1,3}$ is epi.

α is a monomorphism: let M^n be a (singular) G-manifold such that $\alpha(M^n) \sim \emptyset$ in $\Omega_n(X, A; \rho_3)$. Then M^n determines an element $[M]_{\rho_2} \in \Omega_n(X, A; \rho_2)$ such that $t_{2,3}[M]_{\rho_2} = 0$. Since $t_{2,3}$ is a monomorphism, we deduce that there is a ρ_2 bordism $W : M^n \overset{\rho_2}{\sim} \emptyset$. Finally we observe that $\rho_2 \hookrightarrow \gamma$ is a linked embedding of resolutions and therefore W provides the required bordism of M^n to zero in $\tilde{\Omega}_n(X, A; G)$.

The proof of the theorem is now complete.

We are now able to state the theorem asserting the possibility of making bordism with coefficients in a short linked resolution $\rho \xrightarrow{\varepsilon} G$ depend functorially on G.

Theorem 3.7. (a) *There exists a (graded) functor* $\Omega_*(X, A; G) : \mathcal{P} \times \mathcal{A}b \to \mathcal{A}b_*$ *which associates to each pair* $(X, A; G)$ *the graded abelian group* $\Omega_*(X, A; \rho_G)$ *where* ρ_G *is a fixed linked presentation of* G; *to every morphism* $(f, \phi) : (X, A; G) \to (X', A', G')$ *the graded homomorphism* $(f_*, t_{G'}^{-1} \phi_* t_G) : \Omega_*(X, A; \rho_G) \to \Omega_*(X, A; \rho_{G'})$.

(b) *Functors corresponding to different choices of* ρ_G *are naturally equivalent.*

The result follows immediately from Theorem 3.6 and Corollary 3.5.

With the notations of the theorem, we define $\Omega_*(-; G)$, the p.l. oriented bordism with coefficient group G, by $\Omega_*(X; G) = \Omega_*(X; \rho_G)$.

Corollary 3.8. For every pair X, A, every $n \geq 0$ and every abelian group G, there is a short exact sequence

$$0 \to G \otimes \Omega_n(X, A) \to \Omega_n(X, A, G) \to \mathrm{Tor}(G, \Omega_{n-1}(X, A)) \to 0$$

which is natural in (X, A) and in G.

We are now able to say something about the splitting of the universal coefficient sequence associated to $\Omega_*(-; G)$. Precisely, since the sequence is natural on the category $\mathfrak{Ab}$, Hilton [2; Theorem 3.2], gives us the following

Corollary 3.9. For every abelian group G, the universal-coefficient sequence of $\Omega_n(-; G)$ is pure.

From algebra we deduce:

Corollary 3.10. For every pair (X, A), abelian group G and integer $n \geq 0$ such that $\mathrm{Tor}(\Omega_{n-1}(X, A), G)$ is a direct sum of cyclic groups, the universal coefficient sequence

$$0 \to \Omega_n(X, A) \otimes G \to \Omega_n(X, A, G) \to \mathrm{Tor}(\Omega_{n-1}(X, A), G) \to 0$$

splits.

The class of examples of splitting considered by the previous corollary is quite vast, because it includes the following cases:

(a) any G finitely generated

(b) any G such that its torsion subgroup has finite exponent

(c) any $\Omega_{n-1}(X, A)$ such that its torsion subgroup has finite exponent.

Remark 3.11. As we have pointed out earlier, the definition of (ρ, n)-manifold makes sense in the case of ρ being any linked chain complex (not necessarily a resolution) and there is an associated bordism theory $\Omega_*(-;\rho)$. Some of the facts about morphisms, that we have established in this section, hold in the case of chain complexes. In particular, if $\tau : \rho \to \rho'$ is a linked chain map, ρ is a linked chain complex and ρ' is a linked resolution, then the proof of Proposition 3.1 applies to give an associated morphism $\tau_* : \Omega_*(-;\rho) \to \Omega_*(-;\rho')$.

The above treatment of functoriality can be summarized as follows. For every abelian group G, two functors $\mathcal{P} \to \mathcal{A}b_*$ have been set up, namely $\Omega_*(X, A; \rho_G)$ and $\tilde{\Omega}_*(X, A; G)$. They have different features: the former is readily seen to be a generalized homology theory; while the latter is natural on the category of abelian groups. Theorem 3.6 establishes a natural equivalence t_G between the two functors, which proves at the same time that $\Omega_*(X, A; \rho_G)$ is natural on $\mathcal{A}b$ and that $\tilde{\Omega}_*(X, A; G)$ is a homology theory.

In the following we may use whichever of the equivalent functors $\Omega_*(X, A; \rho_G)$, $\Omega_*(X, A; \rho_i)$, $\tilde{\Omega}(X, A; G)$ is more appropriate to the context $(i = 1, 2;\ \rho_i =$ any i-canonical resolution of G).

4. PRODUCTS

If G, G' are abelian groups, let ρ be a linked resolution

$$0 \to F_1 \xrightarrow{\phi} F_0 \xrightarrow{\varepsilon} G \to 0$$

with

$$B^0 = G = \{g_1, g_2, \dots\}$$

$$B^1 = \{r_1, r_2, \dots\}$$

and ρ' defined similarly. Then $\rho \otimes \rho'$ is the augmented chain complex $\{F'', \phi''\}$

$$0 \to F_1 \otimes F_1' \xrightarrow{\phi''_2} F_0 \otimes F_1' \oplus F_1 \otimes F_0' \xrightarrow{\phi''_1} F_0 \otimes F_0' \xrightarrow{\varepsilon''} G \otimes G' \to 0$$

where

$$\phi''_2(r \otimes r') = \phi(r) \otimes r' - r \otimes \phi'(r')$$
$$\phi''_1(g \otimes r') = g \otimes \phi'(r')$$
$$\phi''_1(r \otimes g') = \phi(r) \otimes g'$$
$$\varepsilon'' = \varepsilon \otimes \varepsilon'$$

$\rho \otimes \rho'$ is based by means of B^i, B'^i $(i = 0, 1)$; it is structured in dimension one by the structures of ρ and ρ'; in dimension two we assign to $r \otimes r' \in B'^2$ the word $w(r \otimes r') = w(r) \otimes r' - r \otimes w(r')$. For now we do not fix a cancellation rule.

Let M be a (ρ, i)-manifold with singularities SM; M' a (ρ', j)-manifold with singularities SM'. Form the cartesian product $M \times M'$. It has three intrinsic strata given by

$$(M - SM) \times (M' - SM')$$
$$SM \times (M' - SM') \cup (M - SM) \times SM'$$
$$SM \times SM'$$

On each stratum we can put labels via the tensor product, i.e. if V is a component labelled by x : V' labelled by x', $V \times V'$ is labelled by $x \otimes x'$.

We show that $M \times M'$, with such additional structure, is a $(\rho \otimes \rho', i + j)$-manifold. The first and the second stratum are easily seen to be $\rho \otimes \rho'$-manifolds of the appropriate dimensions and we are going to examine the third stratum. For simplicity assume SM, SM' constantly labelled by r, r' respectively, so that $SM \times SM'$ is labelled by $r \otimes r' \in B''^2$. The basic link of $SM \times SM'$ in $M \times M'$ is topologically the join $L'' = L(r, \rho) * L(r', \rho')$. L" with the structure induced by $M \times M'$ is a $(\rho \otimes \rho', 1)$-manifold because $M \times M' - SM \times SM'$ is a $(\rho \otimes \rho')$-manifold and there is a product structure around $SM \times SM'$. The zero-dimensional stratum of L" is isomorphic to $L(r, \rho) \otimes r' \cup r \otimes -L(r', \rho')$ where $\otimes$ is meant to act on the labels. But this represents the word $w(r \otimes r')$. Thus L" is a $(\rho \otimes \rho', 1)$-manifold in which the zero-dimensional stratum represents $w(r \otimes r')$ and the 1-dimensional stratum is a union of disks. Therefore L" gives a unique cancellation rule to be assigned to $r \otimes r'$ in order to have $L'' = L(r \otimes r', \rho \otimes \rho')$.

$M \times M'$ is then a $(\rho \otimes \rho', i+j)$-manifold. See Figs. 12 and 13.

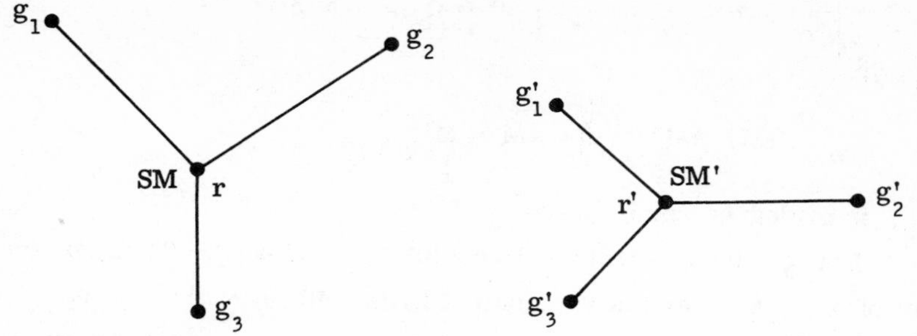
g_1
g_2
g_3
SM
r
g'_1
g'_2
g'_3
SM'
r'

Fig. 12

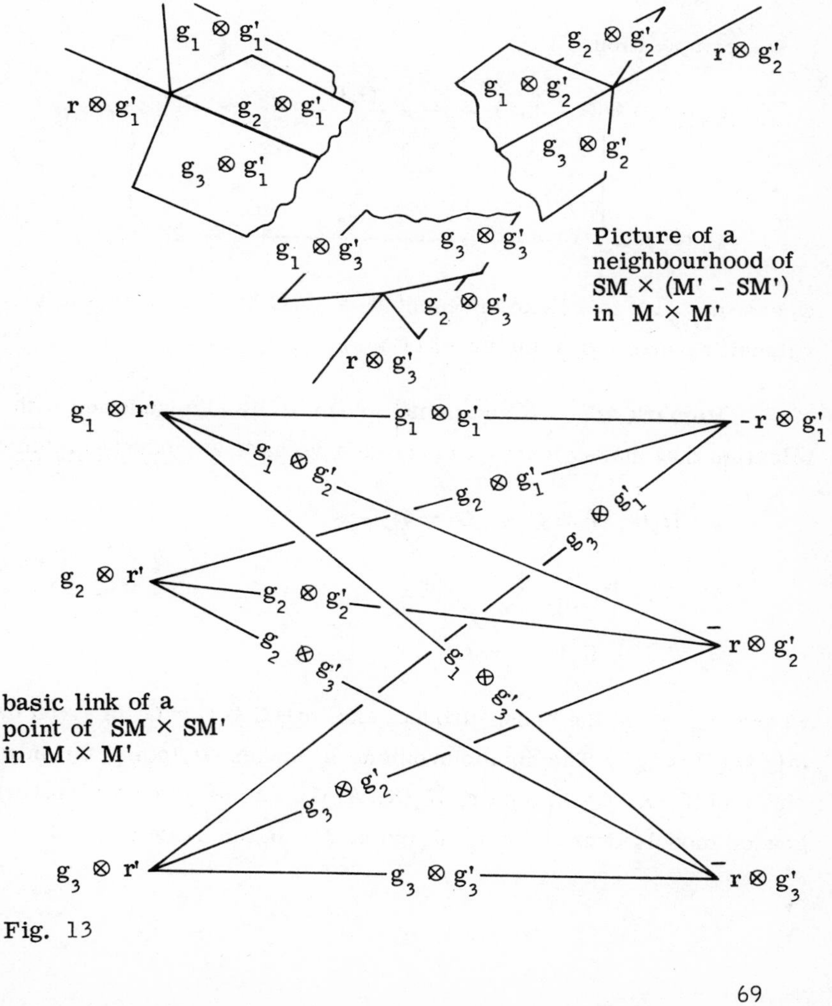
Picture of a neighbourhood of SM × (M' - SM') in M × M'
basic link of a point of SM × SM' in M × M'

Fig. 13

We can now define a homomorphism

$$\times_{\rho,\rho'} : \Omega_*(-;\rho) \otimes \Omega_*(-;\rho') \to \Omega_*(-;\rho \otimes \rho')$$

by

$$\times_{\rho,\rho'}([M]_\rho \otimes [M']_{\rho'}) = [M \times M']_{\rho \otimes \rho'}$$

$\times_{\rho,\rho'}$ is of degree zero.

Let ρ_3 be a 3-canonical resolution of $G \otimes G'$. Then, by the proof of 3.3, there exists a canonical lifting $\widetilde{id} : \rho \otimes \rho' \to \rho_3$ of $id : G \otimes G' \to G \otimes G'$. Therefore we can define a cross-product homomorphism:

$$\times_{G,G'} : \Omega_*(-;G) \otimes \Omega_*(-;G') \to \Omega_*(-;G \otimes G')$$

by the composition

$$\begin{array}{ccc}
\Omega_*(-;\rho) \otimes \Omega_*(-;\rho') & \xrightarrow{\quad \times_{G,G'} \quad} & \Omega_*(-;\rho_{G \otimes G'}) \\
\Big\downarrow {\scriptstyle \times_{\rho,\rho'}} & & \Big\uparrow {\scriptstyle t^{-1}_{1,3}} \\
\Omega_*(-;\rho \otimes \rho') & \xrightarrow{\quad \widetilde{id}_* \quad} & \Omega_*(-;\rho_3)
\end{array}$$

where $\rho_{G \otimes G'}$ is a linked presentation of $G \otimes G'$ and $\widetilde{id}_*$ is the usual relabelling map (as in the proof of 3.6).

Remark 4.1. If an abelian group G is also endowed with a multiplication that makes it into a ring, then we have a product homomorphism

$$\begin{array}{ccc}
\mu : \Omega_*(-;G) \otimes \Omega_*(-;G) & \longrightarrow & \Omega_*(-;G) \\
{\scriptstyle \times_{G,G}} \searrow & & \nearrow {\scriptstyle m_*} \\
& \Omega_*(-;G \otimes G) &
\end{array}$$

where $\times_{G,G}$ is the cross-product and $m : G \otimes G \to G$ is given by: $m(g \otimes g') = gg'$. The homomorphism μ makes $\Omega_*(\text{point}; G)$ into a ring and if (X, A) is a pair, $\Omega_*(X, A; G)$ can be given a structure of graded module over the ring $\Omega_*(\text{point}; G)$ in the usual way.

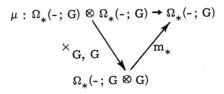

5. THE BOCKSTEIN SEQUENCE

Theorem 5.1. On the category of short exact sequences of abelian groups

$$0 \to G' \xrightarrow{\phi} G \xrightarrow{\psi} G'' \to 0$$

there is a natural connecting homomorphism

$$\beta : \tilde{\Omega}_*(-; G'') \to \tilde{\Omega}_*(-; G')$$

of degree -1 and a natural long exact sequence

$$\cdots \xrightarrow{\beta} \tilde{\Omega}_n(-; G') \xrightarrow{\phi_*} \tilde{\Omega}_n(-; G) \xrightarrow{\psi_*} \tilde{\Omega}_n(-; G'') \xrightarrow{\beta} \cdots$$

Proof. For the sake of clarity of exposition we prove the theorem under the assumptions: $G' \subset G$ and $(X, A) = (\text{point}, \emptyset)$.

Realize the exact sequence of abelian groups by the (not necessarily exact) sequence of canonical resolutions and linked maps

$$\begin{array}{ccccccc} & & \Gamma'_2 & \hookrightarrow & \Gamma_2 & \to & \Gamma''_2 & & \\ & & \downarrow & & \downarrow & & \downarrow & & \\ & & \Gamma'_1 & \hookrightarrow & \Gamma_1 & \to & \Gamma''_1 & & \\ & & \downarrow & & \downarrow & & \downarrow & & \\ & & \Gamma'_0 & \hookrightarrow & \Gamma_0 & \to & \Gamma''_0 & & \\ & & \downarrow & & \downarrow & & \downarrow & & \\ 0 & \to & G' & \hookrightarrow & G & \to & G'' & \to & 0 \end{array}$$

(1) Definition of β

Let M^n be a G''-manifold. Suppose that the singularities of M have only one connected component $V \otimes r''$, with $r'' = g''_1 + \dots + g''_t$ a relation in G''. We relabel V by an element of G' as follows. Choose $g_1, \dots, g_t \in G$ such that $\psi(g_i) = g''_i$ and $g_i = g_j \Longleftrightarrow g''_i = g''_j$. Then $\psi(g_1 + \dots + g_t) = g''_1 + \dots + g''_t = 0$ in G''. Therefore by exactness $g' = \sum_i g_i$ is an element of G'. We relabel V by g' and get a $(G', n-1)$-manifold $V \otimes g'$.

Now suppose $\bar{g}_1, \ldots, \bar{g}_t$ is another lifting of $g_1'', \ldots, g_t''$ as above, giving a (G', n-1)-manifold $V \otimes \bar{g}'$. We show that $V \otimes g'$ and $V \otimes \bar{g}'$ are bordant as (G', n-1)-manifolds. Relabel $M \setminus V$ by changing g_i'' into $g_i' = g_i - \bar{g}_i \in G'$, $i = 1, \ldots, t$. The sum $r' = (g_1' + \ldots + g_t') - (g' - \bar{g}')$ is a relation in G'. Therefore take a labelled copy $V \otimes r'$ of V and form the polyhedron $W = (V \otimes r') \times L(r', G') \cup M$, where $L(r', G')$ is the link generated by r' in G' and the union is taken along the common part $V \times \text{cone}(g_1' + \ldots + g_t')$ (see Fig. 14). W is a (G', n)-manifold and provides the required bordism between $V \otimes g'$ and $V \otimes \bar{g}'$.

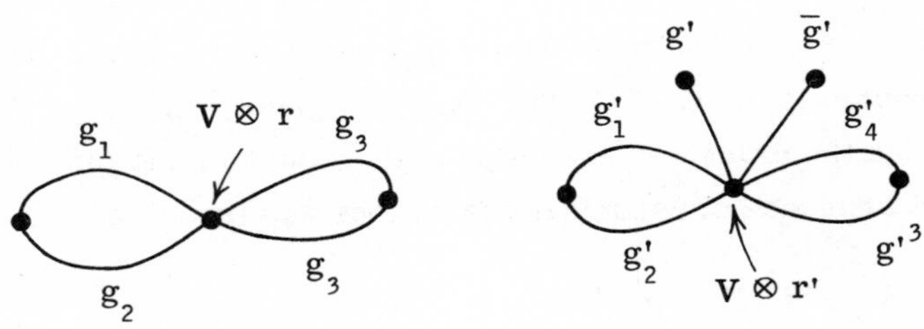

Fig. 14

If the singular set, S(M), of M has more than one component, the relabelling construction can be performed componentwise and one gets a (G', n-1)-manifold, $\beta(M)$, whose bordism class is independent of the various choices.

Next we show that $\beta(M)$ depends only on the bordism class of M, i.e. if $M \overset{G''}{\sim} \emptyset$, then $\beta(M) \overset{G'}{\sim} \emptyset$. Let W be a (G", n+1)-manifold with $\partial W = M$. If $V \subset W \setminus SW$ is a component labelled by $g'' \in G''$, choose $g \in G$ such that $\psi(g) = g''$ and relabel V by g. Let T^n be a component of the n-stratum of W. The sum in G of the new labels on the sheets coming into T^n is an element g' of G'. We relabel T by g'. Finally, if T^{n-1} is a component of the (n-1)-stratum of W and $T_1^n \otimes g_1', \ldots, T_s^n \otimes g_s'$ are the sheets merging into T^{n-1}, then $r' = \sum_{i=1}^{s} g_i'$ is a relation in G', because T^{n-1} was originally labelled by a 'relation amongst relations' of G''. We label T^{n-1} by r'. If $\beta(W)$ denotes SW with the relabelling described above, then $\beta(W)$ provides the required G'-bordism $\beta(M) \sim \emptyset$.

Now we are entitled to define

$$\beta : \tilde{\Omega}_n(-; G'') \to \tilde{\Omega}_{n-1}(-; G')$$

by

$$\beta([M]_{G''}) = [\beta(M)]_{G'} .$$

(2) **Exactness at** $\tilde{\Omega}_n(-; G)$

(a) $\psi_* \phi_* = 0$. If M' is a (G', n)-manifold, then $M'' = \psi\phi(M')$ is a $(0, n)$-manifold. Therefore $M'' \sim \emptyset$ by the proof of the universal-coefficient theorem.

(b) $\operatorname{Ker} \psi_* \subset \operatorname{Im} \phi_*$. Let M^n be a G-manifold and W'' a $(G'', n+1)$-manifold with $\partial W'' = \psi(M)$. We show how to modify W'' in order to get a G-bordism between M^n and a G'-manifold M'^n.

Relabel each component of the $(n+1)$-stratum of W'' by elements of G, obtained from the G''-labels through a lifting $G \overset{\psi}{\underset{\leftarrow - -}{\to}} G''$ such that

(a) The relabelled n-stratum of $\partial W''$ coincides with the n-stratum of M^n.

(b) If two components are labelled by the same element of G'', the corresponding liftings coincide.

Let V be a component of the n-stratum of W'' and $g_1, \ldots, g_v$ the new G-labels around V; $g' = \sum_{i=1}^{v} g_i$ is an element of G'. Attach a new sheet $(V \times I) \otimes g'$ to V iff $g' \neq 0$ and label V by the G-relation $r = \sum_i g_i - g'$ (see Fig. 15). Now let $V \otimes r$, $\overline{V} \otimes \overline{r}$, $\overline{\overline{V}} \otimes \overline{\overline{r}}$, ... be the components of the relabelled n-stratum merging into a component T of the (n-1)-stratum. The corresponding new sheets which have been inserted, namely $(V \times I) \otimes g'$, $(\overline{V} \times I) \otimes \overline{g}'$, $(\overline{\overline{V}} \otimes I) \times \overline{\overline{g}}'$, ..., are, by construction, such that $r' = g' + \overline{g}' + \overline{\overline{g}}' + \ldots$ is a relation in G'. Therefore we can glue them to one another along a new n-dimensional sheet $(T \times I) \otimes r'$. The resulting polyhedron W provides the required G-bordism.

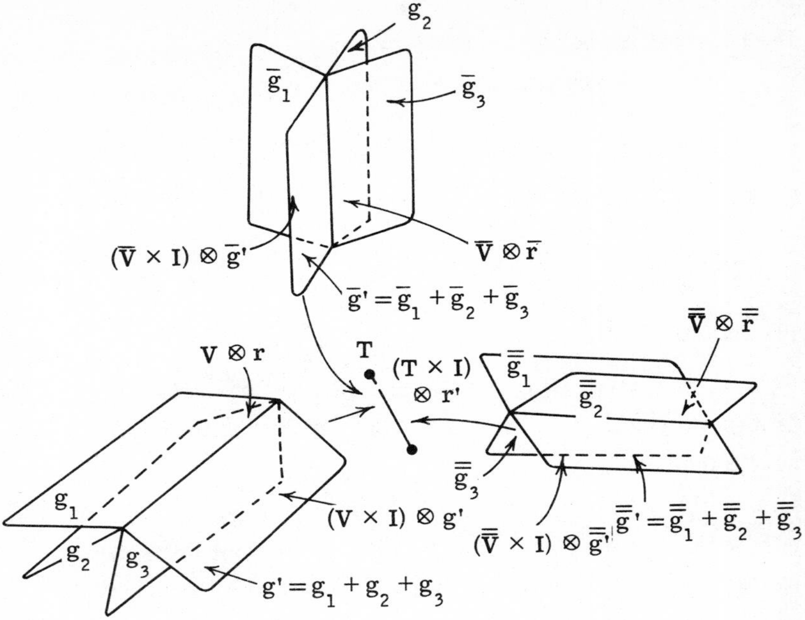

Fig. 15

(3) Exactness at $\tilde{\Omega}_n(-; G'')$

(a) $\beta\psi_* = 0$. Let M^n be a G-manifold. According to the definition of β, we have $\beta\psi(M) = \bigcup_V V \otimes 0$, where V varies over the set of components of $S(M)$. But $V \otimes 0 \sim \emptyset$ by a trivial G'-bordism.

(b) $\operatorname{Ker} \beta \subset \operatorname{Im} \psi_*$. Let M''^n be a G''-manifold. For the sake of simplicity let us assume that the singularities of M'' have only one component $V \otimes r''$; $r'' = \sum_i g''_i$. Then $\beta(M'') = V \otimes g'$, where $g' = \sum_i g_i$, $\psi g_i = g''_i$. By assumption there is a G'-bordism $W' : \beta(M'') \sim \emptyset$. We construct a G-manifold M^n as follows.

(i) Since $\beta(M'')$ has no singularities we can assume that W' has singularities in codimension one at most, because otherwise we solve the codimension-two stratum as in the proof of the universal-coefficient theorem.

(ii) In M" we replace each g''_i by g_i and $V \otimes r''$ by $V \otimes r$, where $r = g' - \sum_i g_i$.

(iii) We attach W' to the relabelled M" identifying $\partial W'$ with $V \otimes r$.

It is readily checked that the resulting labelled polyhedron M^n is a (closed) G-manifold such that $\psi(M) \overset{G''}{\sim} M''$. See Fig. 16

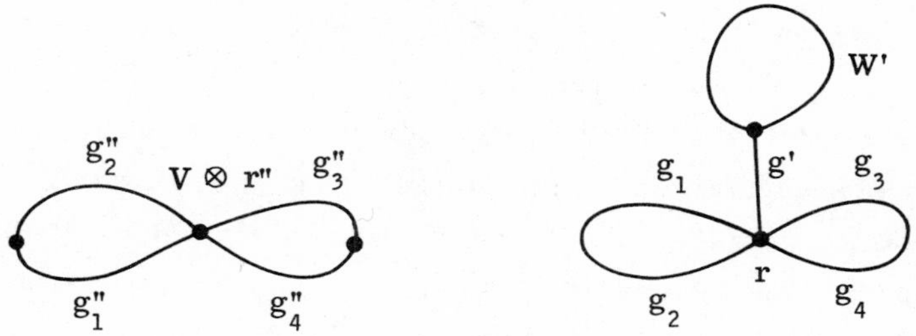

Fig. 16

(4) Exactness at $\tilde{\Omega}_n(-; G')$

(a) $\phi_* \beta = 0$. Let M''^{n+1} be a G"-manifold with connected singularities $V \otimes r''$, $r'' = \sum_i g''_i$. Then $\beta(M'') = V \otimes g'$ where $g' = \sum_i g_i$ and $\psi g_i = g''_i$. We construct a G-bordism $W : \beta(M'') \sim \emptyset$ as follows.

(i) We relabel M by changing each g''_i into g_i and r'' into $r = \sum_i g_i - g'$.

(ii) We attach a new sheet $(V \times I) \otimes g'$ to the relabelled M" along the singularities $V \otimes r$. See Fig. 17

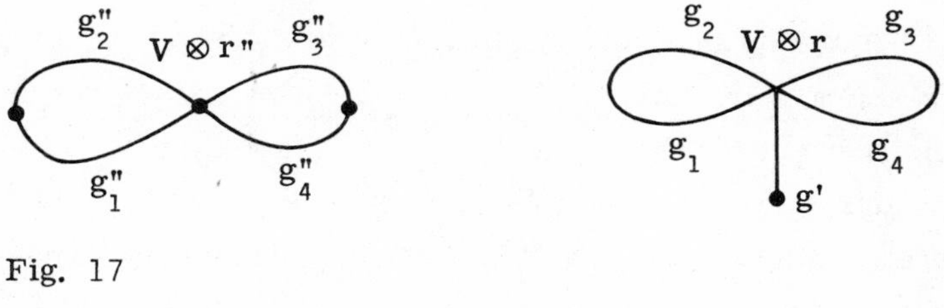

Fig. 17

(b) $\operatorname{Ker} \phi_* \subset \operatorname{Im} \beta$. Let M'^n be a G'-manifold and $W : M'^n \sim \emptyset$ a G-bordism. From W we get a G"-manifold W" of dimension $(n + 1)$ as follows.

(i) We remove from W all the strata which are labelled by elements of G' or by relations or by 'relations amongst relations'.

(ii) We relabel the remaining strata according to the map ψ. The resulting object $\overline{W}"$ is a (G", n+1)-manifold with singularities in codimension ≤ 2 and it is closed because $\partial W = M'$ has been removed in step (i).

(iii) We get W" by re-solving the codimension-two singularities of $\overline{W}"$ up to a bordism which has singularities in codimension ≤ 3.

Now we show that $\beta(W")$ is G'-bordant to M'. Let Q" be the singular part of the bordism used in (iii). Q" has at most two sheets; the non-singular one is labelled by relations in G" and the singular one is labelled by 'relations amongst relations'; $\beta(Q")$ (constructed as in the proof that the Bockstein is well defined) realizes a G'-bordism between $\beta(\overline{W}")$ and $\beta(W")$. Therefore we only need to provide a G'-bordism N' between $\beta(\overline{W}")$ and the original M'. To this purpose we reconsider the G-bordism W and remove from it all the top dimensional strata which are not labelled by elements of G'. The resulting object W_0 is not a G'-manifold in general. We show how to make W_0 into the required G'-bordism by inserting new sheets.

(i) Let $V \otimes r$ be a component of the n-dimensional stratum of W, with $r = g'_1 + \ldots + g'_p + g_1 + \ldots + g_q$ ($g'_i \in G'$; $g_j \in G - G'$). Then $V \otimes g' \subset \beta(\overline{W}")$ where $g' = g_1 + \ldots + g_q$. Therefore we attach a sheet $(V \times I) \otimes g'$ to W_0 along V and change the label r into $r' = g'_1 + \ldots + g'_p + g'$, which is now a relation in G'.

(ii) Let $V \otimes r$, $\overline{V} \times \overline{r}$, $\overline{\overline{V}} \otimes \overline{\overline{r}}$, ... be components merging into a component $T \otimes \tilde{r}$ of the (n-1)-stratum, where $\tilde{r} = r + \overline{r} + \overline{\overline{r}} + \ldots$ is a relation amongst relations in G. The corresponding new sheets which have been inserted, namely $(V \times I) \otimes g'$, $(\overline{V} \times I) \otimes \overline{g}'$, $(\overline{\overline{V}} \times I) \otimes \overline{\overline{g}}'$, ..., are, by construction, such that $r' = g' + \overline{g}' + \overline{\overline{g}}' + \ldots$ is a relation in G'. Therefore we can glue them together along the n-dimensional sheet $(T \times I) \otimes r'$. The resulting polyhedron provides the required G'-bordism $N' : \beta(\overline{W}") \sim M'$.

The proof of exactness is now complete and naturality is clear.

6. BORDISM WITH COEFFICIENTS IN AN R-MODULE

In order to define coefficients in an R-module we need the following additivity lemma.

Lemma 6.1. If $\alpha, \beta : G \to G'$ are abelian-group homomorphisms, then $(\alpha + \beta)_* = \alpha_* + \beta_* : \tilde{\Omega}_*(-; G) \to \tilde{\Omega}_*(-; G')$.

Proof. Consider the chain map $\tilde{\psi} = \tilde{\alpha} + \tilde{\beta} - \widetilde{(\alpha + \beta)}$ where $\tilde{\alpha}, \tilde{\beta}, \widetilde{\alpha + \beta}$ are the canonical liftings of $\alpha, \beta, \alpha + \beta$. If $[M] \in \tilde{\Omega}_n(-; G)$, put $\tilde{\psi}(M) = \tilde{\alpha}(M) + \tilde{\beta}(M) - \widetilde{(\alpha + \beta)}(M)$. Then $\tilde{\psi}(M)$ is a (G', n)-manifold and we only need to prove that $\tilde{\psi}(M) \sim \emptyset$ in $\tilde{\Omega}_n(-; G')$. But $\tilde{\psi}$ is a lifting of the zero map $0 : G \to G'$. Therefore there exists a chain homotopy $D : \tilde{\psi} \simeq 0$.

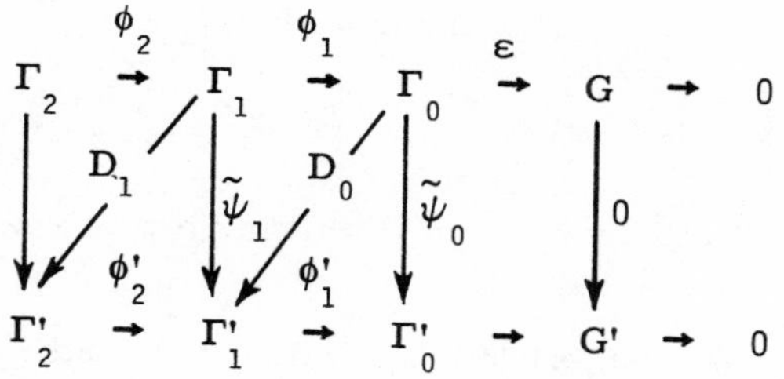

so that $\tilde{\psi}_1 = D_0\phi_1 + \phi'_2 D_1$

$\tilde{\psi}_0 = \phi'_1 D_0.$

By definition the singularities of $\tilde{\psi}(M)$ are given by $\tilde{\psi}(SM)$. Consider the induced diagram

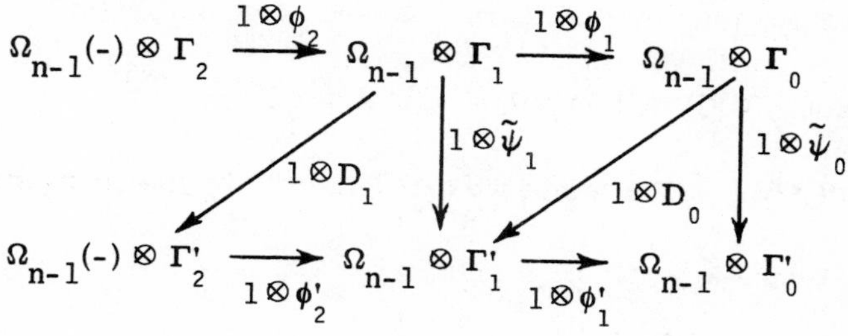

where $1 \otimes D$ is now a homotopy of $1 \otimes \tilde{\psi}$ to zero. As we know, SM represents an element in $\Omega_{n-1}(-) \otimes \Gamma_1$. So we have

$$\tilde{\psi}(SM) = (1 \otimes \tilde{\psi}_1)(SM) = (1 \otimes D_0) \circ (1 \otimes \phi)(SM) + (1 \otimes \phi') \circ (1 \otimes D_1)(SM).$$

But $(1 \otimes \phi)(SM) \sim \emptyset$ because $[SM] = \mathrm{Ker}(1 \otimes \phi_1)$. Therefore $\tilde{\psi}(SM) = (1 \otimes \phi')(W')$, where $W' = (1 \otimes D_1)(SM)$. By the proof of the universal-coefficient theorem this is sufficient to ensure that the singularities $\tilde{\psi}(SM)$ of $\tilde{\psi}(M)$ can be resolved by a bordism of (G', n)-manifolds. So, using the homotopy D, we have eliminated the singular stratum of $\tilde{\psi}(M)$. Let V' be the resulting (G', n)-manifold. If V'_0 is a component labelled by $g'_0 \in G'$, then there exists $g_0 \in G$ such that $\tilde{\psi}_0(g_0) = g'_0$, because the process of resolving the singularities in $\psi(M)$ does not change the labelling of the top dimensional stratum. Therefore the element of $\Omega_n(-) \otimes \Gamma'_0$ represented by V' is the image of some $[V] \in \Omega_n(-) \otimes \Gamma_0$ through $1 \otimes \tilde{\psi}_0$. Then again we have

$$V' = \tilde{\psi}(V) = (1 \otimes \phi')(1 \otimes D_0)(V)$$

so that V' may be borded to $\emptyset$ by a $(G', n+1)$-manifold with singularities given by $(1 \otimes D_0)(V)$.

We now turn to the main object of this section, i.e. putting coefficients in an R-module. In the following R will be a commutative ring with unit.

If G is an R-module, let $\Omega_*(-; G)$ be bordism with coefficients in the underlying abelian group G; $\Omega_*(-; G)$ has a natural R-module structure. In fact, we must exhibit a ring homomorphism $\sigma : R \to \mathrm{Hom}_{\mathbf{Z}}(\Omega_*(-; G), \Omega_*(-; G))$. The above additivity lemma, together with functoriality, tells us that there is a ring homomorphism

$$\sigma' : \mathrm{Hom}_{\mathbf{Z}}(G, G) \to \mathrm{Hom}_{\mathbf{Z}}(\Omega_*(-; G), \Omega_*(-; G))$$

defined by $\sigma'(f) = f_*$. Therefore we can define σ by the composition

$$\begin{array}{ccc} R & \xrightarrow{\sigma} & \mathrm{Hom}_{\mathbf{Z}}(\Omega_*(-; G), \Omega_*(-; G)) \\ & {\scriptstyle \sigma''}\searrow \quad \nearrow{\scriptstyle \sigma'} & \\ & \mathrm{Hom}_{\mathbf{Z}}(G, G) & \end{array}$$

where σ'' is the R-module structure of G.

The pair $\{\Omega_*(-; G), \sigma\}$ is '<u>bordism with coefficients in the R-module</u> G'. The structure σ will be dropped from the notations.

If $f : G \to G'$ is an R-homomorphism, then for every $r \in R$ we have commutative diagrams

$$\begin{array}{ccc} G & \xrightarrow{f} & G' \\ \downarrow r & & \downarrow r \\ G & \xrightarrow{f} & G' \end{array} \qquad\qquad \begin{array}{ccc} \Omega_*(-; G) & \xrightarrow{f_*} & \Omega_*(-; G') \\ \sigma(r) \downarrow & \text{functoriality} & \downarrow \sigma(r) \\ \Omega_*(-; G) & \xrightarrow{f_*} & \Omega_*(-; G') \end{array}$$

Hence $\Omega_*(-; G)$ is a functor on the category of R-modules and R-homomorphism. From the naturality of the Bockstein sequence for abelian groups, it follows that there is a functorial Bockstein sequence in the category of R-modules. Summing up, we have the following:

(a) $\Omega_*(-; G)$ is a functor on the category of R-modules

(b) $\Omega_*(-; G)$ is additive

(c) For every short exact sequence of R-modules, there is an associated functorial Bockstein sequence.

Properties (a), (b), (c) form the hypothesis of Dold's Universal-coefficient theorem [1].

Therefore we deduce that there is a spectral sequence running

$$E^2_{p,q} = \mathrm{Tor}_p(\Omega_q(-; R), G) \underset{p}{\Rightarrow} \Omega_*(-; G)$$

This completes the discussion of the case of R-modules as coefficients. In later chapters we shall only deal with abelian groups; but it is understood that everything we say continues to work in the category of R-modules.

REFERENCES FOR CHAPTER III

[1] A. Dold. Universelle Koeffizienten. <u>Math. Zeitschrift,</u> 80 (1962/3).

[2] P. J. Hilton. Putting coefficients into a cohomology theory. <u>Konikl. Nederl. Akademie van Weterschappen (Amsterdam), Proceedings, Series A,</u> 73 No. 3 and <u>Indag. Math.</u> 30 No. 3, (1970), 196-216.

[3] P. J. Hilton and A. Deleanu. On the splitting of universal coefficient sequences. Aarhus Univ., Algebraic topology Vol. I, (1970), 180-201.

[4] J. Morgan and P. Sullivan. The transversality characteristic class and linking cycles in surgery theory. Ann. of Math. 99 (1974), 463-544.

IV · Geometric theories

In this chapter we extend the notion of a geometric homology and cohomology (mock bundle) theory by allowing

(1) singularities

(2) labellings

(3) restrictions on normal bundles.

The final notion of a 'geometric theory' is in fact sufficiently general to include all theories (this being the main result of Chapter VII). A further extension, to equivariant theories, will be covered in Chapter V.

In the present chapter, we also deal with coefficients in an arbitrary geometric theory. A geometric theory with coefficients is itself an example of a geometric theory and it is thus possible to introduce coefficients repeatedly!

The chapter is organised as follows. In §1 we extend the treatment of coefficients in the last chapter to cover oriented mock bundles and in §§2 and 3 we deal with singularities and restrictions on the normal bundle. In §§4 and 5 we give interesting examples of geometric theories, including Sullivan's description of K-theory [11] and some theories which represent (ordinary) Z_p-homology. Finally §6 deals with coefficients in the general theory.

1. COBORDISM WITH COEFFICIENTS

We now combine Chapters II and III to give a geometric description of cobordism with coefficients. It is first necessary to introduce oriented mock bundles (the theory dual to oriented bordism). We give here the simplest definition of orientation, an alternative definition will be given in §2.

Suppose M^n, V^{n-1} are oriented manifolds with $V \subset \partial M$. Then we define the incidence number $\varepsilon(V, M) = \pm 1$ by comparing the orientation of V with that induced on V from M (the induced orientation of ∂M is defined by taking the inward normal last); $\varepsilon(V, M) = +1$ if these orientations agree and -1 if they disagree. An oriented cell complex K is a cell complex in which each cell is oriented and then we have the incidence number $\varepsilon(\tau, \sigma)$ defined for $\tau^{n-1} < \sigma^n \in K$.

An oriented mock bundle is a mock bundle ξ/K in which each block is oriented, K is oriented and such that, for each $\tau^{n-1} < \sigma^n \in K$, we have $\varepsilon(\xi(\tau), \xi(\sigma)) = \varepsilon(\tau, \sigma)$. We leave the reader to check that the theory of oriented mock bundles enjoys all the properties of the unoriented theory in Chapter II (the Thom isomorphism theorem holds for oriented bundles and Poincaré duality for oriented manifolds) - more general arguments will in fact be given in §2. This theory will be denoted $\Omega^*(\, , \,)$ and the dual bordism theory $\Omega_*(\, , \,)$.

Now let G be an Abelian group and ρ a structured resolution of G. We define the mock bundle theory $\Omega^*(\, , \, ; \rho)$ by using ρ-manifolds in place of ordinary manifolds. More precisely, a (ρ, q)-mock bundle ξ/K is a polyhedron $E(\xi)$ with projection $p : E(\xi) \to K$ such that, for each $\sigma^i \in K$, $p^{-1}(\sigma)$ is a $(\rho, q+i)$-manifold with boundary $p^{-1}(\dot{\sigma})$, called the block over σ and denoted $\xi(\sigma)$, and such that $\varepsilon(\xi(\tau), \xi(\sigma)) = \varepsilon(\tau, \sigma)$ for each $\tau < \sigma \in K$. Note, $\varepsilon(V, M)$ is defined for ρ-manifolds V^{n-1}, M^n only when either V or $-V \subset \partial M$ as ρ-manifolds (i.e. the inclusion respects the labellings, orientations and extra structure), then $\varepsilon(V, M) = +1$ in the first case and -1 in the second. -V denotes the ρ-manifold obtained from V by reversing all the orientations.

It follows from the arguments in Chapters I and II that $\Omega^*(, \, ; \rho)$ is a cohomology theory, dual to the theory $\Omega_*(\, , \, ; \rho)$ defined in Chapter III, and from the arguments in III §4 that $\Omega^*(P, Q; -)$ determines a functor on the category of abelian groups.

We will leave most of the details to the reader and make some remarks about some of the more delicate situations:

Remarks 1.1. 1. If ξ/K is a (ρ, q)-mock bundle and $|K|$ is an (oriented) n-manifold, then $E(\xi)$ is a $(\rho, n+q)$-manifold. The proof of this is identical to the proof in Chapter II - the required extra structure all comes automatically!

2. In order to prove Poincaré duality, one needs Cohen's transversality theorem in its full generality, i.e. if $f : J \to K$ is simplicial, then $f^{-1}(\tilde{A})$ is collared in $f^{-1}(A^*)$ for each $A \in K$. Here A^* is the dual cone of A in K with base $\tilde{A}$. From this theorem it follows that, if $f : E \to K$ is a simplicial map, E is a $(\rho, n+q)$-manifold and $|K|$ is an n-manifold, then the inverse image of a dual cell in K cuts the singularities of E transversally, so that $f : E \to K$ can be made into the projection of a (ρ, q)-mock bundle.

3. A discussion completely analogous to that of III §4 can be carried out. In particular there are functors $\Omega^*(\; , \; ; G)$ natural on the category of abelian groups and there is a universal coefficient sequence

$$0 \to \Omega^q(\; , \;) \otimes G \to \Omega^q(\; , \; ; G) \to \mathrm{Tor}(\Omega^{q-1}(\; , \;), G) \to 0$$

also natural on the category of abelian groups.

4. If $\phi : G \otimes G' \to G''$ is a pairing, the *cup product* $\Omega^q(\; ; G) \otimes \Omega^r(\; ; G') \to \Omega^{q+r}(\; ; G'')$ and the *cap product* $\Omega^q(\; ; G) \otimes \Omega_r(\; ; G') \to \Omega_{q+r}(\; ; G'')$ are defined using the usual pull-back construction and the cross product defined in Chapter III.

2. RESTRICTIONS ON NORMAL BUNDLES

In this section we consider geometric (co)-homology theories which can be obtained from pl (co)-bordism by restricting the normal bundles of the manifolds considered. We sketch the case of cobordism. Details for the bordism case may be found in [13; Chapter II].

Let $E(\xi) \xrightarrow{p} K$ be a mock bundle projection, then we can choose an embedding $i : E(\xi) \to K \times R^\infty$ so that $p = \pi_1 \circ i$, we then have a stable normal block bundle $\nu_\xi / E(\xi)$ on $E(\xi)$ in $K \times R^\infty$. There is a classifying bundle map

$$\begin{array}{ccc} E(\nu_\xi) & \xrightarrow{\hat{\eta}_\xi} & E(\gamma) \\ \cup & & \cup \\ E(\xi) & \xrightarrow[\eta_\xi]{} & B\widetilde{PL} \end{array}$$

where $\gamma/B\widetilde{PL}$ is the classifying bundle for stable block bundles.

Now suppose that we have a space X and a fibration $f : X \to B\widetilde{PL}$. Then an (X, f)-mock bundle is a mock bundle ξ together with a stable normal block bundle ν_{ξ}, a classifying bundle map $(\hat{\eta}_{\xi}, \eta_{\xi})$ and a lift of η_{ξ} in X:

$$\begin{array}{ccc} & & X \\ & \overset{q_{\xi}}{\nearrow} & \downarrow f \\ E(\xi) & \xrightarrow{\eta_{\xi}} & B\widetilde{PL} \end{array}$$

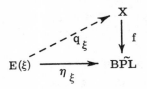

The theory of (X, f)-mock bundles is set up in exactly the same way as the theory of bundles. In order to have products one needs in addition a commutative diagram

$$\begin{array}{ccc} X \times X & \xrightarrow{m} & X \\ \downarrow f \times f & & \downarrow f \\ B\widetilde{PL} \times B\widetilde{PL} & \xrightarrow{\oplus} & B\widetilde{PL} \end{array} \qquad (2.1)$$

where $\oplus$ is the map given by Whitney sum. Using diagram (2. 1) external products can be defined by $q_{\xi \times \eta} = m \circ (q_{\xi} \times q_{\eta})$. Similarly cap products are defined with the corresponding bordism theory (maps of (X, f)-manifolds into the space) and the proof of Poincaré duality (for (X, f)-manifolds) needs little change. The proof of the Thom isomorphism theorem for bundles with stable lifts in X can also be readily modified.

Examples 2. 2. 1. Oriented theory. $X = BS\widetilde{PL}$ and f is the natural map. This theory has products. See also the alternative description given in §1.

2. Smooth cobordism. X has the homotopy type of BO and $f : X \to B\widetilde{PL}$ is defined using $\widetilde{PD}$ as in [7; §0]. This again is a theory with products.

3. Pl spin cobordism. X is the double cover of $BS\widetilde{PL}$ and f is the covering map. Again we have products.

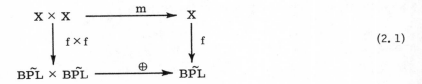

4. <u>Stable cohomotopy.</u> X is contractible. Again we have products. Poincaré duality holds for π-manifolds.

5. <u>Labelling.</u> Let S be a discrete set and let $X = \widetilde{BPL} \times S$ and f the projection. Then a connected (X, f)-manifold is just a manifold labelled by an element of S. Any function $S \times S \to S$ gives this theory products. See also the remarks at the end of the next section.

3. SINGULARITIES

Our treatment of singularities is similar to that worked out by Cooke and Sullivan (unpublished) or to be found in Stone [9].

Suppose we are given a class $\mathcal{L}_n$ of (n-1)-polyhedra (closed under pl isomorphism). Then a closed $\mathcal{L}_n$-manifold is a polyhedron M each of whose links lies in $\mathcal{L}_n$. A theory of manifolds-with-singularity consists of a class $\mathcal{L}_n$ for each $n = 0, 1, \ldots$ which satisfies:

1. each member of $\mathcal{L}_n$ is a closed $\mathcal{L}_{n-1}$-manifold
2. $S\mathcal{L}_{n-1} \subset \mathcal{L}_n$ (i.e. the suspension of an (n-1)-link is an n-link)
3. $C\mathcal{L}_{n-1} \cap \mathcal{L}_n = \emptyset$ (i.e. the cone on an (n-1)-link is never an n-link).

Then an $\mathcal{L}$-manifold with boundary is a polyhedron whose links lie either in $\mathcal{L}_n$ or $C\mathcal{L}_{n-1}$. Then the boundary consists of points whose links lie in the latter class, and is itself a closed $\mathcal{L}_{n-1}$-manifold. Moreover the boundary is locally collared (since its links are cones) and hence collared [8; 2.25].

Notice that axiom 3 is necessary to ensure that the boundary is well-defined. Axiom 2 ensures that if M is an $\mathcal{L}_{n-1}$-manifold then $M \times I$ is an $\mathcal{L}_n$-manifold with boundary. Axiom 1 implies that a regular neighbourhood of a polyhedron in an $\mathcal{L}_n$-manifold is itself an $\mathcal{L}_n$-manifold with boundary.

At this point we can remark that a manifold with singularities has all the geometry of an ordinary manifold which was used in setting up bordism and cobordism (mock bundles) and we get homology and cohomology theories $T^*_{\mathcal{L}}(\ ,\ )$, $T^{\mathcal{L}}_*(\ ,\ )$. Moreover the proofs of the Poincaré duality and Thom isomorphism theorems are unaltered. Note however, see below, that products are not defined in general (but cap product with the fundamental class of a manifold (amalgamation) is always defined).

Products

Suppose M and N are closed $\mathcal{L}$-manifolds then $M \times N$ is in general not an $\mathcal{L}$-manifold. However it is one if we have:

4. $\mathcal{L}_n * \mathcal{L}_q \subset \mathcal{L}_{n+q}$ (i.e. the join of two links is again a link).

Then, with axiom 4, we have cup and cap products. More generally if $\mathcal{L}$, $\mathcal{M}$ and $\mathcal{N}$ are three theories and $\mathcal{L} * \mathcal{M} \subset \mathcal{N}$ then we have cup and cap products from $\mathcal{L}$ and $\mathcal{M}$ theory to $\mathcal{N}$ theory. For example, if $\mathcal{S}$ is ordinary bordism theory (i.e. $\mathcal{S}_n = \{(n-1)\text{-sphere}\}$) then $\mathcal{L} * \mathcal{S} \subset \mathcal{L}$ by axiom 2, so that, as remarked above, cup and cap products with bordism or cobordism classes are always defined.

Basic links

A subset $\mathcal{B}$ of $\mathcal{L} = \cup \mathcal{L}_n$ is basic if no link in $\mathcal{B}$ is a suspension and each link in $\mathcal{L}$ is isomorphic to a suspension of a link in $\mathcal{B}$.

Examples 3.1. 1. $\mathcal{L}_n$ is the class of (n-1)-spheres. As mentioned above, this is ordinary bordism theory. The set of basic links here is $\{\emptyset\}$.

2. Basic links are $\emptyset \in \mathcal{L}_0$ and (n points) $\in \mathcal{L}_1$. Thus $\mathcal{L}_0 = \{\emptyset\}$, $\mathcal{L}_1 = \{X \mid X \cong S^0$ or (n points)$\}$, $\mathcal{L}_q = \{X \mid X \cong S^{q-1}$ or $S^{q-2} *$ (n points)$\}$, $q \geq 1$.

This theory is 'twisted $\mathbf{Z}_n$-manifolds'. A manifold in the theory is either locally an ordinary manifold or like $R^{n-1} \times C$ (n points). This theory can be made into 'coefficients $\mathbf{Z}_n$' by adding orientations and an untwisted neighbourhood for the singularity (see Chapter III, Example 1.1(2)). The twisted theory is interesting in connection with representing $\mathbf{Z}_n$-homology (see §5).

3. $\mathcal{L}_0 = \{\emptyset\}$, $\mathcal{L}_1 = \{X \mid X \cong S^0\}$, $\mathcal{L}_n$ is all closed $\mathcal{L}_{n-1}$-manifolds.

This theory is 'ordinary' homology with coefficients $\mathbf{Z}_2$. To obtain coefficients $\mathbf{Z}$ one needs to orient the top stratum. We can think of this theory as obtained from bordism by killing all manifolds except the point. (§4 contains details of killing.)

To combine $\mathcal{L}$-theory with the restriction on the normal bundle of the last section, it is necessary to use the notion of 'normal block bundle system' as in Stone [9]. The resulting theories then enjoy all the usual properties - the class of bundles and manifolds for which a theory has Thom and Poincaré isomorphisms depends on the stable restrictions imposed on the normal bundles. Rather than attempt a formal analysis of this general setting, we will give several examples in subsequent sections, which should make the general properties of these theories clear. We already have the examples, in Chapter III, of coefficients (all the structure of a manifold with coefficients ρ is included in 'restriction on the normal block bundle system') and in Chapter VII, we will give a family of examples, generated by the killing process of §4, below, which include all homology theories.

Finally we remark that we have now arrived at the general notion of a geometric theory, since 'labelling' is included in 'restriction on normal bundle', see Example 2.2(5).

4. KILLING AND K-THEORY

In this section we give the general description of 'killing' an element of a theory and apply it to give a geometric description of connected K-theory at odd primes due essentially to Sullivan [11]. See also Baas [15]. Killing is defined in the following generality:

1. U and V are geometric theories.
2. M is a closed (V, n)-manifold.
3. There is a natural way of regarding $W \times M$ as a V-manifold, for each U-manifold W (e.g. by relabelling or forgetting some structure).

Then the theory $V/U \times M$ is defined by considering polyhedra P with a two stage stratification $P \supset S(P)$ and extra structure such that:

1. $P - S(P)$ is a (V, q)-manifold.
2. S(P) is a (U, q-n-1)-manifold.
3. There is a regular neighbourhood N of S(P) in P and a pl isomorphism $h : N \to S(P) \times C(M)$, which carries S(P) by the identity to $S(P) \times$ (cone pt.).

4. h is an isomorphism of (V, q)-manifolds off S(P) (where $S(P) \times (C(M) - \text{(cone pt.)})$ is regarded as a V-manifold by part 3 of the data).

P is then called a closed $(V/U \times M, q)$-manifold. There is an obvious notion of $V/U \times M$-manifold with boundary and hence we have geometric homology and cohomology theories $(V/U \times M)_*$ and $(V/U \times M)^*$

Notation. If U = V then we collapse the notation to V/M.

Proposition 4.1. <u>There are long exact sequences</u>

$$\to U_q(X, A) \xrightarrow{\chi} V_{q+n}(X, A) \xrightarrow{\iota} (V/U \times M)_{q+n}(X, A) \xrightarrow{\sigma} U_{q-1}(X, A).$$

$$\to U^q(P, Q) \xrightarrow{\chi} V^{q+n}(P, Q) \xrightarrow{\iota} (V/U \times M)^{q+n}(P, Q) \xrightarrow{\sigma} U^{q-1}(P, Q).$$

<u>in which the homomorphisms are defined as follows.</u> χ <u>is multiplication by</u> M <u>followed by the identification of part 3 of the data.</u> ι <u>is the identity on representatives.</u> σ <u>restricts to the second stratum [i.e.</u> $\sigma(P \supset S(P), f) = (S(P), f|S(P))$ etc.].

Remark 4.2. There is also a notion of killing a whole family $\{M_i\}$ of elements simultaneously. In this definition $S(P) = \bigcup_i (S(P)_i)$ and the neighbourhood of $S(P)_i$ satisfies the conditions of the definition but with M_i replacing M. This is then a generalisation of the killing used in Chapter III to define coefficients. 4.1 becomes sequences like:

$$\to \bigoplus_i U_q \xrightarrow{\chi} V_{q+n} \xrightarrow{\iota} (V/U \times M_i)_{q+n} \xrightarrow{\sigma} \bigoplus_i U_{q-1} \cdots$$

which the reader can check is a generalisation of the Universal Coefficient Sequence. C.f. Remark 6.1.

Proof of 4.1. (Compare the proof of the universal coefficient formula.) The spaces X, A, P, Q play no role in the proof, so we ignore them.

<u>$\iota\chi = 0$.</u> Let W be a U-manifold then $W \times M$ bounds the $V/U \times M$-manifold $W \times C(M)$.

<u>Ker $\iota \subset$ Im χ.</u> Let W be a V-manifold bordant to $\emptyset$ by $V/U \times M$-

bordism W_1. Then $M \times S(W_1)$ is bordant to W by the V-bordism W_1 - (nbhd. of $S(W_1)$).

$\sigma\iota = 0.$ A V-manifold has no second stratum.

Ker $\sigma \subset$ Im ι. Let $(W, S(W))$ be a $V/U \times M$-manifold such that $S(W)$ bounds the U-manifold B. Form the product $B \times C(M)$ and attach it to $W \times \{1\}$ in $W \times I$ by the identity on $S(W) \times C(M)$. This constructs a $V/U \times M$-bordism of $W = W \times \{0\}$ to a V-manifold.

$\chi\sigma = 0.$ If $(W, S(W))$ is a $V/U \times M$-manifold then $S(W) \times M$ bounds W - (nbhd. of $S(W)$).

Ker $\chi \subset$ Im σ. If W is a U-manifold such that $W \times M$ bounds the V-manifold W', then we can glue $W \times C(M)$ to W' along $W \times M$ to form a $V/U \times M$-manifold with second stratum W.

Corollary 4.3. Suppose V is a ring theory and $[M]$ is not a zero-divisor in $V_*(\text{pt.})$ (i.e. multiplication by $[M]$ is injective) then

$$(V/M)_*(\text{pt.}) \cong \frac{V_*(\text{pt.})}{\text{ideal generated by } [M]} .$$

Proof. Consider the sequence with $X = \text{pt.}$ $A = \emptyset$.

From now on we will, for notational simplicity, deal only with the homology theories. Exactly similar constructions will hold for the cohomology theories.

Let $\Omega_*^{SO}(\ , ; \mathbf{Z}[\frac{1}{2}])$ be the theory 'smooth bordism with $\mathbf{Z}[\frac{1}{2}]$ coefficients' defined by considering ρ-manifolds, where ρ is a fixed resolution of $\mathbf{Z}[\frac{1}{2}]$, with a reduction to SO of the stable normal bundle of each intrinsic stratum (see the last two sections). By the universal coefficient sequence, this theory is isomorphic to $\Omega_*^{SO}(\ ,) \otimes \mathbf{Z}[\frac{1}{2}]$ the localisation of Ω_*^{SO} at odd primes.

From results of Wall [14] we know that all the torsion in Ω_*^{SO} is 2-torsion and hence that $\Omega_*^{SO}(\text{pt.} ; \mathbf{Z}[\frac{1}{2}])$ is a free polynomial algebra on generators $[M_1^4]$, $[M_2]$, ..., moreover we can take index$(M_1) = 1$ and index$(M_i) = 0$, $i > 1$: take $M_1 = \mathbf{CP}_2$ and to obtain index $(M_i)=0$ subtract an appropriate number of copies of $(\mathbf{CP}_2)^j$. Now define theories J^i, $i = 1, 2, \ldots$ as follows:

$$J^1 = \Omega_*^{SO}(\ ,\ ; \mathbf{Z}[\tfrac{1}{2}]).\quad J^2 = J^1/M_2,\ \text{and inductively}\ J^i = J^{i-1}/M_i.$$

Finally let

$$J = \lim\{J^1 \to J^2 \to J^3 \to \dots\}.$$

Thus J is the geometric theory obtained from smooth bordism by introducing coefficients in $\mathbf{Z}[\frac{1}{2}]$ and then killing all the free generators except CP_2. Note that by repeated use of 4.3 we have:

Proposition 4.4. $J_*(\mathrm{pt.}) \cong \mathbf{Z}[\frac{1}{2}][t]$ <u>where</u> t <u>has dimension</u> 4, <u>and is represented geometrically by</u> CP_2 <u>labelled by</u> 1.

Now let K denote the theory $ko_*(\ ,\ ) \otimes \mathbf{Z}[\frac{1}{2}]$, i.e. the localization of real connected K-theory at odd primes.

Theorem 4.5. <u>There is a natural equivalence of theories</u>

$$\psi : K \to J.$$

Proof. Sullivan [10], using a method similar to Conner and Floyd [3], has constructed a natural transformation

$$s : \Omega_*^{SO}(\ ,\ ) \otimes \mathbf{Z}[\tfrac{1}{2}] \to K_*(\ ,\ )$$

such that $s(\mathrm{pt.})$ maps $[M^n]$ to 0 if $n \neq 4k$ and to $\mathrm{index}(M)\, t^k$, where t is the generator of $K_*(\mathrm{pt.}) \cong \mathbf{Z}[\frac{1}{2}][t]$, if $n = 4k$. He also proves that s induces an isomorphism

$$\Omega_*^{SO}(\ ,\ ) \otimes_{\Omega_*^{SO}} \mathbf{Z}[\tfrac{1}{2}][t] \cong K_*(\ ,\ )$$

where $\Omega_*^{SO} = \Omega_*^{SO}(\mathrm{pt.})$ acts on $\mathbf{Z}[\frac{1}{2}][t]$ by

$$\Omega_*^{SO} \xrightarrow{\otimes 1} \Omega_*^{SO} \otimes \mathbf{Z}[\tfrac{1}{2}] \xrightarrow{s} K_*(\mathrm{pt.}) \cong \mathbf{Z}[\tfrac{1}{2}][t].$$

We will construct a natural transformation ψ in the commutative diagram

$$
\begin{array}{ccc}
\Omega_*^{SO}(\ ,\ ;\mathbf{Z}[\tfrac{1}{2}]) = \Omega_*^{SO}(\ ,\ )\otimes\mathbf{Z}[\tfrac{1}{2}] & & \\
\swarrow \iota \qquad\qquad \searrow \text{natural projection} & & (4.6) \\
J_*(\ ,\ ) \xleftarrow{\ \psi\ } \Omega_*^{SO}(\ ,\ )\underset{\Omega_*^{SO}}{\otimes} \mathbf{Z}[\tfrac{1}{2}][t] \cong K_*(\ ,\ ) & &
\end{array}
$$

and then it follows from 4.4 that $\psi(\mathrm{pt.})$ is an isomorphism since the class of $\mathbf{CP}_2$ is the generator of both groups. ψ is defined on generators by the formula $\psi((M, f), qt^k) = (qM \times (\mathbf{CP}_2)^k, f \circ \pi_1)$, where $q \in \mathbf{Z}[\frac{1}{2}]$ and qM means M labelled by q. We have to check that ψ is well-defined. The only non-trivial part is that if $[W] \in \Omega_*^{SO}(\mathrm{pt.})$ then $\psi((M \times W^n, f \circ \pi_1), 1) = \psi((M, f), s(W))$. I. e. that

$$(M \times (W - \mathrm{index}(W)(\mathbf{CP}_2)^{n/4}), f \circ \pi_1)$$

is zero in J (here $\mathrm{index}(W) = 0$ if $n \neq 4k$, for brevity of notation). This follows from:

Proposition 4.7. $\iota[W] = 0$ _if_ $\mathrm{index}(W) = 0$. ι _is as in diagram 4.6._

Proof. $[W] = \sum \alpha_i W_i$ in $\Omega_*^{SO}(\ ,\ ; \mathbf{Z}[\frac{1}{2}])$ where $\alpha_i \in \mathbf{Z}[\frac{1}{2}]$ and W_i are monomials in the generators $\mathbf{CP}_2 = M_1, M_2, \ldots$. However, using the product formula for the index, we can read off $\mathrm{index}(W)$ as α_1 where α_1 is the coefficient of $(M_1)^{n/4}$, since all the other M_i have index 0. It follows that $\alpha_1 = 0$, and we can bord W to $\emptyset$ in the theory J by using bordisms like $C(M_2) \times \overline{W}_i$, where $W_i = M_2 \overline{W}_i$.

Remark 4.8. There is a similar geometric description of connected KU-theory given by a similar construction using complex bordism and the Conner-Floyd map [3].

5. MORE EXAMPLES

Example 5.1. _Some theories which represent Z_p-homology._

Let p be prime. Define a theory of singularities by the basic links $\emptyset \in \mathcal{L}_0$, $(p) \in \mathcal{L}_1$, $(p) * (p) \in \mathcal{L}_2, \ldots$ where (p) is a set with p

points in it. We call an $\mathcal{L}$-manifold a p-polyhedron. This is a ring theory (see §3), the ring closure of 'twisted $\mathbf{Z}_p$-bordism' (Example 3.1(2)). An orientation for an n-dimensional p-polyhedron is a generator of $H_n(P; \mathbf{Z}_p) \cong \mathbf{Z}_p$. The theory of oriented p-polyhedra represents $\mathbf{Z}_p$-homology, in other words the natural maps $T_n(\ ) \to H_n(\ ; \mathbf{Z}_p)$ and $T^q(\ ) \to H^q(\ ; \mathbf{Z}_p)$ are onto. This follows from:

Proposition (see [6]). Let U be a connected ring theory with $U_0(\text{pt.}) \cong \mathbf{Z}_p$. Then U represents $\mathbf{Z}_p$-homology if and only if $U^q(L_n) \to H^q(L_n; \mathbf{Z}_p)$ is onto, where L_n is the Lens space S^n/p of arbitrarily high dimension n.

Now the generators of $H^*(L_n; \mathbf{Z}_p)$ are α and β, where $\alpha \in H^2$ is represented by the inclusion of L_{n-2} in L_n, and $\beta \in H^1$ is represented by $L_{n-2} \cup_\partial D^{n-1} \to L_n$, where the disc is glued on by the p-fold cover. Note that the Bockstein of β is α. Both α and β are p-polyhedra and the representation property follows.

We can modify $T_{\mathcal{L}}$ in various ways, still preserving its property of representing $\mathbf{Z}_p$-homology, for example:

1. Make stable restrictions on the normal bundles of the strata. E.g. impose stable orthogonal or unitary structures. Note that α and β have such structures.

2. The normal bundle of one stratum in the others can be restricted. I.e. we can restrict the freedom to 'twist'. The point is that the group used in the construction of β is $\mathbf{Z}_p$ not Σ_p as allowed for in the definition of a p-polyhedron. To make this restriction into a ring theory restriction, we impose the restriction that the group for the normal block bundle of a stratum of codimension $r + q$ in a stratum of codimension r is the wreath product $\mathbf{Z}_p \wr \mathbf{Z}_p \wr \mathbf{Z}_p \dots \wr \mathbf{Z}_p$ (r copies).

Both these modifications are examples of restriction on the normal block bundle system. For more information on the algebra behind p-polyhedra see Bullett [1].

Example 5.2. Euler spaces.

This theory was invented by Akin and Sullivan [16] and has interesting properties. Define link classes by

$$\mathcal{L}_0 = \{\emptyset\},\ \mathcal{L}_1 = \{(q) \mid q \text{ even}\} \text{ and inductively}$$
$$\mathcal{L}_n = \{P \mid P \text{ is a closed } \mathcal{L}_{n-1}\text{-manifold with even Euler characteristic}\}.$$

Then an $\mathcal{L}$-manifold is called an Euler space and can be thought of as a polyhedron with 'even local Euler characteristic'. Note that manifolds are Euler spaces and that Euler spaces form a ring theory. An Euler space has Steifel homology classes, [12, 16], (defined using the combinatorial definition of Whitney et al., see Halperin and Toledo [2]). The triangulation of a complex algebraic variety is an example of an Euler space (Sullivan proves this by a careful induction on dimension using the fact that each stratum is even-dimensional).

Example 5.3. The Casson-Quinn theories.

Finally we mention some examples of geometrically defined (co)-homology theories, which do not fit as described into the pattern of this chapter. These are the theories whose coefficients are the surgery obstructions. For details of the definition see Quinn [5]. A 'manifold' in the theory is a surgery problem with a reference space (corresponding to fundamental group) and a boundary on which the problem is a homotopy equivalence. Of particular interest is the theory $(\Omega^4 G/PL)_*$, which is the Casson-Quinn theory corresponding to $\pi_1 = 0$. Sullivan has shown [10] that, at odd primes, this theory is isomorphic to K-theory, as in §4. This raises the question of whether there is a convenient geometrical description for $(\Omega^4 G/PL)_*$ (or even for G/PL_* itself) at all primes. Also relevant here is the question of whether K-theory has a simpler geometrical representation than that given in §4. Note always that, by Chapter VII, all cohomology theories have some geometric representation.

6. COEFFICIENTS IN A GEOMETRIC THEORY

In this section V denotes a general geometric theory, that is to say, a theory with singularities, labellings and generalised orientations, as in §§2, 3. We will explain how coefficients work for V-bordism. We leave the reader to take care of V-cobordism (V-mock bundles) and to formulate the appropriate Thom isomorphism and duality theorems. This section is modelled on Chapter III, we follow the section headings of Chapter III, explaining where the difficulties lie.

Short resolutions

Let ρ be a short resolution of an abelian group G. A V-manifold with coefficients in ρ can be defined exactly as in III §1 and the theory enjoys all the analogous properties. In particular there is a universal coefficient theorem.

Remark 6.1. Coefficients ρ is an example of killing, as described in §4. To make the notation fit with §4, let $V_1 = V \otimes F_0$ (V-manifolds labelled by elements of B_0) and $U_1 = V \otimes F_1$. The transformation $U_1 \times L(r, \rho) \to V_1$ is given by ignoring the label on the first factor. Then (V, ρ)-theory is the theory obtained from V_1 by killing simultaneously the elements $\{L(r, \rho) | r \in B_1\}$.

Longer resolutions

The description of coefficients in resolutions of length ≤ 4 in Chapter III is again an example of killing (made precise as in 6.1 above). To make this work for a general theory we need to regard $L(b_i, \rho) \times M$ as a (V, ρ)-manifold for each V-manifold M and each link $L(b_i, \rho)$. Now each stratum of $L(b_i, \rho)$ is a disc, so we can regard $L(b_i, \rho) \times M$ as a stratified set with each stratum a V-manifold, and the only possible problem comes from 'restrictions on the normal bundle'. In general this problem is solved by endowing $L(b_i, \rho)$ with the universal restriction, namely framings of each stratum which fit together in a standard way (i.e. $L(b_i, \rho)$ is an object in the theory of framed manifolds with coefficients - stable homotopy with coefficients). For the 0-stratum there is no problem (framing is equivalent to orientation). For the 1-stratum we have to frame each 1-disc extending given framings near the ends. Orientation considerations imply that this is possible but there is non-uniqueness - there are two possible choices for each 1-disc. Finally for the 2-stratum, the non-uniqueness of framings of circles implies that the framing may not be possible. We now make some more precise statements.

Definition 6. 2. Let $\tilde{S}^1$ denote the circle with the non-standard framing. V is a good theory if there is a bordism D of $\tilde{S}^1$ to zero in V such that $M \times D$ is a bordism of $M \times \tilde{S}^1$ to zero for each $[M] \in V_*(\text{pt.})$.

Remarks 6. 3. 1. For ring theories our definition of a good theory coincides with Hilton's, see [4; 1. 9]. For general theories Hilton's definition is equivalent to insisting that $\eta \times \tilde{S}^1$ is cobordant to zero for each V-mock bundle η. This is in fact sufficient to prove Theorem 6. 4(1), but we will not give details.

2. If V is a good theory then we can complete the construction of $L(b_2, \rho)$ - plug in the bordism D of $\tilde{S}^1$ to zero wherever appropriate (in most cases $D = D^2$ and the construction coincides with the old one).

Functoriality

The best result we have, for a general theory, is the following:

Theorem 6. 4. Coefficients in a short resolution of an abelian group gives a notion of coefficients which is functorial
(1) for good theories
(2) on the category of direct sums of free abelian groups and odd torsion groups.

Remarks 6. 5. The universal coefficient sequence is natural (and hence, usually, splits) in exactly the same cases.

Proof of 6. 4. 1. Exactly the same proof as III §4, using Remark 6. 3(2).

2. Step 1. Coefficients are always functorial on the category of free abelian groups.

This is seen as follows. Define $\tilde{\Omega}'_q(\ ; F)$, where F is a free abelian group, by allowing no singularities in the representatives (i. e. labellings only) and codimension 1 singularities only in the bordisms (this requires only the definition of 0-links, which, as seen earlier, always holds). Now $\tilde{\Omega}'_q(\ ; F)$ is the same as $\Omega_q(\ ) \otimes F$ by exactly the argument of III §4, but with all levels of singularities reduced one step.

We leave the reader to check the details here; geometrically all that is required is a construction (not unique) of 1-links, which we gave above.

<u>Step 2.</u> Coefficients are functorial for odd torsion groups.

The idea here is to use the fact that $\tilde{S}^1$ has order two to complete the construction of the 2-links for a 3-canonical resolution. At the final stage we have $\tilde{S}^1$ labelled by $g \in G$ and g has order t, t odd. We have to plug in a bordism of $g\tilde{S}^1$ to zero. Take $\frac{t-1}{2}$ copies of $S^1 \times I$ framed so $\tilde{S}^1$ is at both ends and glue all the copies of $\tilde{S}^1$ together. Finally glue on one copy of $\tilde{S}^1 \times I$ by one end. This constructs the required bordism. The new singularity is labelled by the relation $g + g + \ldots + g$ (t times).

<u>Step 3.</u> Coefficients are functorial on the category of direct sums.

Let $G = F \oplus G_1$, where F is free and G_1 is an odd torsion group. Define $\tilde{\Omega}''_q(\ ; G)$ by using the two definitions given above. Precisely a generator of $\tilde{\Omega}''_q$ is the union of a generator of $\tilde{\Omega}_q(\ ; G_1)$. A 'bordism' is similarly a union of bordisms. Now let $G' = F' \oplus G'_1$ and $h : G \to G'$ a homomorphism. Then h splits as

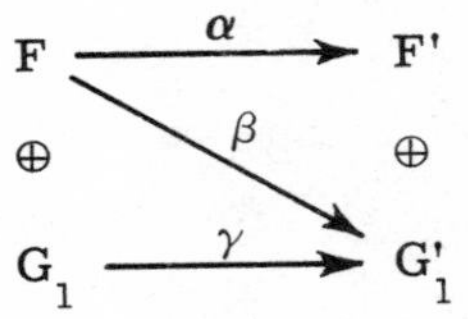

since there is no non-trivial homomorphism $G_1 \to F'$.

This means that we can define $h[M]$ by simply relabelling and we never meet the problem of relabelling an element with singularities of too high a codimension. Similarly bordisms can be relabelled. Thus $\tilde{\Omega}''$ is functorial. That it is isomorphic to the correct group follows from Steps 1 and 2. This completes the proof.

Products, Bockstein sequence and rings of coefficients

The rest of Chapter III goes through with obvious changes in the general case. The constructions are functorial under the same conditions as Theorem 6.4.

REFERENCES FOR CHAPTER IV

[1] S. Bullett. Ph. D. thesis, Warwick University (1973).

[2] S. Halperin and D. Toledo. Stiefel Whitney homology classes. Ann. of Math. 96 (3), (1972), 511-25.

[3] P. E. Conner and L. E. Floyd. The relation between cobordism and K-theory. Springer-Verlag Lecture Notes No. 28.

[4] P. J. Hilton and A. Deleanu. On the splitting of universal coefficient sequences. Aarhus Univ., Algebraic Topology, Vol. I, (1970), 180-201.

[5] F. S. Quinn. A geometric formulation of surgery. Ph. D. Thesis, Princeton (1969).

[6] C. P. Rourke. Representing homology classes. Bull. Lon. Math. Soc., 5 (1973), 257-60.

[7] C. P. Rourke and B. J. Sanderson. Block bundles: III. Ann. of Math., 87, (1968), 431-83.

[8] C. P. Rourke and B. J. Sanderson. Introduction to pl topology. Springer-Verlag, Berlin (1972).

[9] D. A. Stone. Stratified polyhedra. Springer-Verlag lecture notes No. 252.

[10] D. Sullivan. Geometric topology lecture notes. M. I. T. (1970).

[11] D. Sullivan. Geometric topology seminar notes. Princeton (1967).

[12] D. Sullivan. Combinatorial invariants of analytic spaces. (To appear.)

[13] R. E. Strong. Notes on cobordism theory. Princeton U. P. (1968).

[14] C. T. C. Wall. Determination of the cobordism ring. Ann. of Math., 72 (1960), 292-311.

[15] N. A. Baas. On bordism theory of manifolds with singularity. Math. Scand., 33 (1973), 279-302.

[16] E. Akin. Stiefel Whitney homology classes and bordism. Trans. Amer. Math. Soc., 205 (1975), 341-59.

V·Equivariant theories and operations

In §1 of this chapter we describe a further extension of mock bundles, to the equivariant case - the theory dual to equivariant bordism - and in §2 give a general construction which includes power operations and characteristic classes. The remainder of the paper is concerned with the case of $\mathbf{Z}_2$-operations on pl cobordism. In §3 we expound the 'expanded squares' ('expanded' rather than the familiar 'reduced' because of our indexing convention for cohomology) and in §4 we give the relation with tom Dieck's operations [5]. §5 describes the characteristic classes associated to $\mathbf{Z}_2$-block bundles and in §6 we give a result inspired by Quillen [3] which relates the total square of a mock bundle with the transfer of the euler class of its twisted normal bundle. This leads, in some cases, to the familiar connection between characteristic classes and squares. Finally in §7 we give an alternative definition of squares, based on transversatility. This is like the 'internal' definition of the cup product (see II end of §4).

1. EQUIVARIANT MOCK BUNDLES

Let G be a finite group and X a polyhedron. By a G-action on X we mean a (pl) map $G \times X \to X$ satisfying

(i) for all $g_1, g_2 \in G$ and $x \in X$, $g_1(g_2 x) = (g_1 g_2)x$.

(ii) if $e \in G$ is the identity then $ex = x$ for all $x \in X$.

If X, Y are G polyhedra then a map $f : X \to Y$ is a G-map if f commutes with the G-action. We then have the concept of <u>equivariant bordism</u> of X by equivariantly mapping G-manifolds into X and we show below how to define equivariant cobordism via G-mock bundles.

Suppose now G acts on $X = |K|$. Then we say G acts on K provided for each $\sigma \in K$ and $g \in G$, $g\sigma \in K$. The action is <u>good</u> if in addition whenever $g\sigma = \sigma$ we have $g/\sigma = \text{id}$.

Proposition 1.1. Suppose G acts on X. Then there exists K such that $|K| = X$ and there is induced a good action on K.

Proof. It follows easily from definitions that X/G has a *pl* structure so that the quotient map $q : X \to X/G$ is *pl*. Choose $|L| = X/G$, so that for each $g \in G$ the subpolyhedron $q\{x | gx = x\}$ is a subcomplex of L. Then define $\sigma \in K$ if and only if $q(\sigma) \in L$ then K is the desired complex.

A mock bundle ξ/K is a G-mock bundle if there is an action of G on $E(\xi)$ which induces an action on K. Let $T^q_G(K)$ denote the group of G-cobordism classes of G-mock bundles over K. If the action on $E(\xi)$ is good (i. e. if $g\beta_\sigma = \beta_\sigma$ implies $g|\beta_\sigma = \text{id.}$) then the subdivision process can be carried through equivariantly simply by subdividing all blocks in an orbit isomorphically.

It follows from Proposition 1.1 and the subdivision construction that there is a G-homotopy functor T^q_G on G-polyhedra defined by considering only good actions. In fact we have:

Proposition 1.2. There is a natural isomorphism

$$\theta : T^q(X/G) \to T^q_G(X).$$

Proof. Let $[\xi/L] \in T^q(X/G)$, with L as in the proof of 1.1. Then $q^*\xi$ has a natural G-action and we can define $\theta[\xi/L] = [q^*\xi]$. θ is easily proved to be an isomorphism.

From 1.2 it is easily seen that the work of II carries over to the equivariant case when the action is free, for example there are cup and cap products and Thom and Poincaré duality isomorphisms.

If U is a geometric theory then there is also a notion of equivariant U-bordism and equivariant U-mock bundles. We omit details.

We shall see in the next section that the case of a G-mock bundle ξ/K with the action in K not good is extremely interesting as the power operations spring from consider such cases.

2. THE GENERAL CONSTRUCTION AND THE POWER OPERATIONS

Let W be a free G-polyhedron and J a G-complex with fixed point polyhedron $F \subset |J| = X$.

Let U be a geometric theory. We have the following commutative diagram of homomorphisms:

$$\begin{array}{ccc} & \xrightarrow{\Phi_1} & U^q(X \times_G W) \\ & & \downarrow r^* \\ U^q_G(J) & \xrightarrow{\Phi_0} & U^q(F \times W/G) \\ & \overset{\Phi_!}{\dashrightarrow} & \downarrow \pi_! \\ & & U^{q+n}(F) \end{array} \qquad (2.1)$$

Here $\Phi_1(\xi) = \theta^{-1}(\xi \times 1_W)$ and G acts on $X \times W$ by the diagonal action. r^* is restriction and $\pi_!$ is composition with the trivial mock bundle with fibre W/G, when this is defined, that is, when $M \times W/G$ is a U-manifold for each U-manifold M, (e.g. if W/G is a U-manifold and U is a ring theory).

The whole diagram is natural for subdivisions of G-bundles over J, for G inclusions $J_0 \subset J$ and $W_0 \subset W$. Further $\Phi_!$ depends only on the free G-cobordism class of W.

There is a relative version got by replacing J, Q, and F by (J, J_0), (Q, Q_0), and (F, F_0) respectively.

The construction of Φ_0, Φ_1 can be made for more conventional types of bundles, for example vector bundles, spherical fibrations.

Example 2.1. Let u/X be a G-vector bundle with $G = \mathbf{Z}_2$ and let W be the sphere S^n with antipodal action. Then $u|F = u_0 \oplus u_1$ where G acts trivially on the fibres of u_1 and antipodally in the fibres of u_0. It easily follows that $\Phi_0(u) = \pi_1^*(u_0) \otimes \pi_2^*(l_n) + \pi_1^*(u_1)$ where l_n is the canonical line bundle on P_n and π_1 and π_2 are the obvious projections.

Power operations

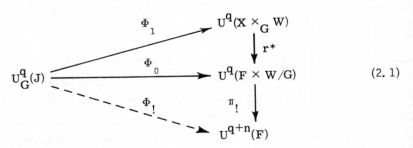

Now let Σ_r denote the symmetric group on r symbols and suppose given a non-trivial homomorphism $\mu : G \to \Sigma_r$. Define

$s : Z^q(K) \to Z^{qr}_G(K^r)$, where Z denotes isomorphism classes of U-mock bundles,

by $s(\xi) = \overbrace{\xi \times \ldots \times \xi}^{r}$ with G action given by permuting factors via μ. s commutes with subdivision of K and can be seen to define <u>external power operations</u>

$$P_1(\mu, W) : U^q(X) \to U^{qr}(X^r \times_G W)$$

$$P_0(\mu, W) : U^q(X) \to U^{qr}(X \times W/G)$$

by $P_i(\mu, W) = \Phi_i \circ s$. When $\pi_!$ is defined (see above) then we have internal operations $P_!(\mu, W) = \Phi_! \circ s : U^q(X) \to U^{qr+n}(X)$.

3. THE EXPANDED SQUARES

Now restrict attention to ordinary pl cobordism (denoted $T^*(,)$ as usual) and, in the construction of the end of the last section, let $G = Z_2 = \Sigma_2$ and let $W = S^n$, the pl n-sphere with antipodal action. We obtain the <u>external and internal expanded squares:</u>

$$\begin{array}{ccc} & \overset{Sq^n_1}{\nearrow} & T^{2q}(X \times X \times_{Z_2} S^n) \\ & & \downarrow \\ T^q(X) & \xrightarrow{Sq^n_0} & T^{2q}(X \times P^n) \\ & & \downarrow \\ & \underset{Sq^n_!}{\searrow} & T^{2q+n}(X) \end{array}$$

for $n = 0, 1, 2, \ldots$.

Remark 3.1. The name 'expanded square' gains more credence from the observation that we can choose representatives ξ^{2r+n} for $Sq^{n+r}_![\xi^r]$ so that

$$E(\xi^{2r}) \subset E(\xi^{2r+1}) \subset \ldots \subset E(\xi^{2r+n}) \subset \ldots ,$$

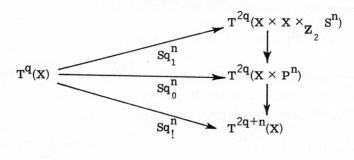

and from definitions we have $[\xi^{2r}] = [\xi^r] \cup [\xi^r]$. A similar remark applies to $Sq_i^n[\xi]$, $i = 0, 1$.

The following lemma is easily proved.

Lemma 3.2. (a) $Sq_0^0[\xi^r] = [\xi^r] \cup [\xi^r]$.

(b) $i^*Sq_1^n[\xi] = [\xi] \times [\xi]$, where $i : X \times X \times pt. \to X \times X \times_{\mathbf{Z}_2} S^n$ is the inclusion induced by the inclusion of a point in S^n.

(c) Let u/K be a block bundle and let tu be its (canonical) Thom class then $Sq_1^n(tu)$ is the (canonical) Thom class of $E(u) \times E(u) \times_{\mathbf{Z}_2} S^n \to X \times X \times_{\mathbf{Z}_2} S^n$.

(d) $Sq_!^{n+m}[M^m] = [M \times M \times_{\mathbf{Z}_2} S^n]$ where M is a closed manifold here regarded as an m-mock bundle over a point.

Proposition 3.3. Sq_1^n, Sq_0^n and $Sq_!^{n+r}$ are homomorphisms.

Proof. It is sufficient to show Sq_1^n is a homomorphism. Abusing the notation we have

$$Sq_1^n(\xi + \eta) = Sq_1^n(\xi) + Sq_1^n(\eta) + ((\xi \times \eta + \eta \times \xi) \times S^n)/\mathbf{Z}_2.$$

The last factor is a composition (of mock bundles)

$$\xi \times \eta \times S^n \to K \times K \times S^n \to (K \times K \times S^n)/\mathbf{Z}_2,$$

but the 0-mock bundle $K \times K \times S^n \to (K \times K \times S^n)/\mathbf{Z}_2$ gives the class $0 \in T^0((K \times K \times S^n)/\mathbf{Z}_2)$. The result follows.

The following is immediate from definitions.

Proposition 3.4. Let $i : ((K \times L) \times (K \times L) \times S^n)/\mathbf{Z}_2 \to (K \times K \times S^n)/\mathbf{Z}_2 \times (L \times L \times S^n)/\mathbf{Z}_2$ be defined by $i[x_0, x_1, y_0, y_1, z] = ([x_0, y_0, z], [x_1, y_1, z])$. Then $Sq_1^n[\xi \times \eta] = i^*(Sq_1^n[\xi] \times Sq_1^n[\eta])$ for any ξ, η.

Corollary 3.5. Sq_1^n and Sq_0^n are ring homomorphisms.

Proof. It is sufficient to show that Sq_1^n is a ring homomorphism but from the formula in 3.4 and the commutative diagram

$$\begin{array}{ccc} ((K \times K) \times (K \times K) \times S^n)/Z_2 & \xrightarrow{\ i\ } & (K\times K\times S^n)/Z_2 \times (K\times K\times S^n)/Z_2 \\ & {\scriptstyle \Delta\times\Delta\times 1}\nwarrow \qquad \nearrow{\scriptstyle \Delta} & \\ & (K \times K \times S^n)/Z_2 & \end{array}$$

we get $Sq_1^n([\xi] \cup [\eta]) = Sq_1^n[\xi] \cup Sq_1^n[\eta]$.

The relation between $Sq_!^i \xi \cup Sq_!^i \eta$ and $Sq_!^k(\xi \cup \eta)$ is illuminated by the following.

Corollary 3. 6. <u>There is a commutative diagram</u>

$$\begin{array}{ccc} E(Sq_!^{n+r}\xi^r \cup Sq_!^{n+s}\eta^s) & \longleftarrow & E(Sq_!^{n+r+s}(\xi \cup \eta)) \\ \downarrow & & \downarrow \\ X \times P_n \times P_n & \xleftarrow{\ 1\times\Delta\ } & X \times P_n \\ & \searrow \quad \swarrow & \\ & X & \end{array}$$

(with the outer maps $E(Sq_!^{n+r}\xi^r \cup Sq_!^{n+s}\eta^s) \to X$ and $E(Sq_!^{n+r+s}(\xi \cup \eta)) \to X$)

<u>in which the square is a pull back.</u>

Proof. This follows from 3. 5 and the diagram

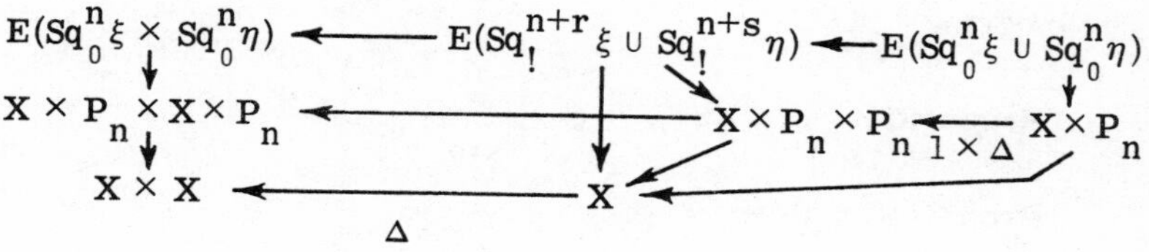

$$\begin{array}{ccccc} E(Sq_0^n\xi \times Sq_0^n\eta) & \longleftarrow & E(Sq_!^{n+r}\xi \cup Sq_!^{n+s}\eta) & \longleftarrow & E(Sq_0^n\xi \cup Sq_0^n\eta) \\ \downarrow & & \downarrow \quad \searrow & & \downarrow \\ X \times P_n \times X \times P_n & \longleftarrow & X\times P_n \times P_n & \xleftarrow{\ 1\times\Delta\ } & X \times P_n \\ \downarrow & & \swarrow & & \\ X \times X & \xleftarrow{\ \Delta\ } & X & \longleftarrow & \end{array}$$

in which each square is a pull back.

Proposition 3. 7. $j^*Sq_0^nSq_0^m[\xi] = Sq_0^mSq_0^n[\xi]$, <u>where</u> $j : Q \times P_n \times P_m \to Q \times P_m \times P_n$ <u>is given by</u> $j(x, y, z) = (x, y, z)$.

Proof. This follows from commutativity in

$$\begin{array}{ccc} (Q\times P_n)\times P_m & \xrightarrow{\ \Delta\times\Delta\ } & ((Q\times Q\times S^n)/Z_2 \times (Q\times Q\times S^n)/Z_2 \times S^m \times S^m)/Z_2 \\ \cong \downarrow j & & \cong \downarrow k \\ (Q\times P_m)\times P_n & \xrightarrow{\ \Delta\times\Delta\ } & ((Q\times Q\times S^m)/Z_2 \times (Q\times Q\times S^m)/Z_2 \times S^n \times S^n)/Z_2 \end{array}$$

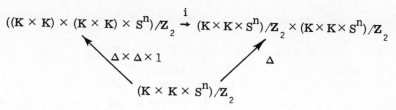

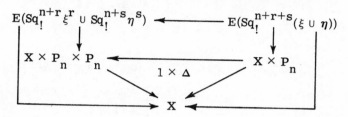

where k shuffles the spheres and each Δ is a suitable diagonal map.

4. RELATIONS WITH TOM DIECK'S OPERATIONS

In [5] tom Dieck defines operations in the smooth case analogous to our Sq_0 and Sq_1. That the definitions agree in the smooth case follows from 3.2.c. We now look at the relationship between our internal operation $Sq_!^i$ and tom Dieck's internal operation. By virtue of the Thom isomorphism, see e.g. [1], one readily proves

$$T^*(X \times P_n) \cong T^*(X) \otimes T^*(P_n)$$

and

$$T^*(P_n) \cong T^*[x]/\langle x^{n+1}\rangle$$

where x is the euler class of the canonical line bundle l_n/P_n and by direct construction we have $p_x : P_{n-1} \to P_n$, the usual inclusion, and the projection of x^i is the inclusion $P_{n-i} \to P_n$. Thus

$$Sq_0^n(\xi^q) = \sum_{i=0}^{n} R^{q+i}\xi^q \otimes x^i$$

and the R^{q+i} are tom Dieck's internal operations, with a change of sign in the indexing.

Proposition 4.1. $Sq_!^{n+q}(\xi) = \sum_{i=0}^{n} [P_{n-i}]R^{-q-i}\xi^q$.

Proof. Consider

$$T^*(X) \otimes T^*(P_n) \xrightarrow{\cong} T^*(X \times P_n) \xrightarrow{p_!} T^*(X).$$

Then $p_! x^i$ has projection $X \times P_{n-i} \to X$ and therefore $p_! x^i = [P_{n-i}].1$. Further if ξ/X has projection $p_\xi : E(\xi) \to X$ then $\xi \otimes x^i$ has projection $E(\xi) \times P_{n-i} \to X \times P_n$ and composing with $X \times P_n \to X$ we see that $p_! \xi \otimes x^i = [P_{n-i}]\xi$. The result follows.

Now let $\mu : T^*(-) \to H^*(-; \mathbf{Z}_2)$ be the Steenrod map (the identity on representatives, see IV 3.1(3)). Let Sq^i be the usual Steenrod squar (with a change of sign).

Proposition 4.2. (i) $\mu Sq^i_!(\xi) = Sq^i \mu(\xi)$, <u>where on the right we have the usual Steenrod operation.</u>

(ii) $\mu Sq^i_!(\xi) = \mu R^i(\xi)$.

Proof. (i) follows from the axiomatic description of Sq^i and from (ii), which comes from the fact that

$$\mu[P_{n-i}] = 0 \text{ unless } n = i.$$

5. CHARACTERISTIC CLASSES

In this section we present a special case of the construction of §1. Let u/K be a block bundle with involution $f : E(u) \to E(u)$ satisfying $fp^{-1}\sigma = p^{-1}\sigma$ and $|K| = \{x : f(x) = x\}$. Recall that the inclusion $X = |K| \subset E(u)$ is the projection of the Thom class tu and by virtue of the involution we have $[tu] \in T_G^{-s}(E(u), \dot{E}(u))$, $G = \mathbf{Z}_2$. Now we may apply the construction of §1 to get

$$\underline{W}_n(u) = \Phi_0[tu] \in T^{-s}(X \times P_n),$$

and

$$W^!_{n-s}(u) = \Phi_![tu] \in T^{n-s}(X).$$

The classes $W^!_{-s}(u), W^!_{-s+1}(u), \ldots, W^!_{-s+n}(u), \ldots$ are the $\mathbf{Z}_2$ <u>characteristic classes</u> of (u, f).

Denote by $u \otimes l_n$ the block bundle $E(u) \times_{\mathbf{Z}_2} S^n \to X \times P_n$. In the case that u is a vector bundle with antipodal action then $u \otimes l_n$ coincides with the usual tensor product with the canonical line bundle.

From definitions we have:

Proposition 5.1. $\underline{W}_n(u)$ <u>is the euler class</u> $e(u \otimes l_n)$, <u>and</u> $W^!_{-s}(u^s) = e(u)$.

To get some geometric insight into the meaning of the $W^!_k(u)$ consider the diagram

$$\begin{array}{ccc}
E(\underline{W}_n(u)) & \dashrightarrow & X \times P_n \\
\downarrow & & \downarrow p \\
X \times P_n & \subset & (E(u) \times S^n)/\mathbf{Z}_2 \\
\downarrow \pi_1 & & \downarrow \\
X & \subset & E(u)/\mathbf{Z}_2
\end{array}$$

where p is the projection of $\Phi_1(tu)$ after subdivision so that $X \times P_n \subset (E(u) \times S^n)/\mathbf{Z}_2$ appears as a subcomplex. From the diagram we see that we may regard $E(\underline{W}_n(u))$, after dividing out the Z_2 action, as $f^{-1}(X)$ where f is an equivariant approximation to $X \times S^n \to X \subset E(u)$ which is transverse to X. Such a map f may be produced by using the pl transversatility theorem and an induction over the cells of $K \times S^n$, where S^n has a suitable equivariant cell structure. This gives an alternative definition of the characteristic classes.

From the alternative definition we have

Proposition 5.2. <u>Let</u> ε^s/K <u>be the trivial block bundle then</u> $W_r^!(\varepsilon^s)$ <u>is the class of the projection</u> $K \times P_r \to K$.

Now suppose v^{r+1}/K and u^s/K are block bundles with involution and that there is an equivariant isomorphism $u^s \oplus v^{r+1} \cong \varepsilon^{r+s+1}$ where ε is the trivial block bundle with standard involution and the involution on $u^s \oplus v^{r+1}$ is induced from $u^s \oplus v^{r+1} \cong p_u^* v \mid E(u)$. Then we have $K \times S^{r+s} \subset E(\varepsilon^{r+s+1})$ and composing isomorphisms and the projection into $E(u)$ we get an equivariant transverse map

$$f : K \times S^{r+s} \to E(u) \text{ with } f^{-1}|K| = P(v),$$

where $P(v)$ denotes the mock bundle $E(\dot{v})/Z_2 \to K$. We have then:

Proposition 5.3. <u>In the above situation,</u> $W_r^!(u^s)$ <u>is the class of</u> $P(v) \to |K|$.

6. SQUARES AND EULER CLASSES

The purpose of this section is to prove Theorem 6.1 below and derive consequences. The result was inspired by Theorem 3.12 of Quillen [3].

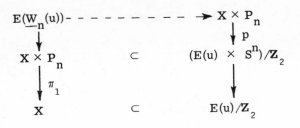

Let u/X be a pl bundle with fibre some euclidean space, and suppose $\Delta : X \times P_n \to (X \times X \times S^n)/Z_2$ is given by $\Delta(x, [y]) = [(x, x, y)]$. Then there is a bundle inclusion $u \times P_n \to \Delta^*(u \times u \times S^n/Z_2)$. Define $u^+/X \times P_n$ to be the complementary block bundle. The existence and uniqueness of u^+ follows from 5.1 of [4].

Now suppose given a mock bundle ξ^q/K with $|K| = X$ and $E_\xi \subset E_u$ where u is a pl bundle with fibre $\mathbf{R}^{q+m}$ and for each $\sigma \in K$, $p_\xi^{-1}(\mathring{\sigma}) = p_u^{-1}(\mathring{\sigma}) \cap E_\xi$ and further suppose given a pl normal bundle v/E_ξ for the inclusion so that $p_v^{-1} p_\xi^{-1}(\mathring{\sigma}) = p_u^{-1}(\mathring{\sigma}) \cap p_v^{-1}(E_\xi)$. In this situation we say ξ is in u with normal bundle v.

Theorem 6.1. *Suppose* ξ^q/X *is in* u^{q+m} *with normal bundle* v^m. *Then* $e(u^+)Sq_0^n(\xi) = p_! e(v^+)$, *where* $p : E_\xi \times P_n \to X \times P_n$ *is given by* $p(x, y) = (p_\xi(x), y)$.

We need the following generalisation of the clean intersection formula, 3.3 of [3].

Lemma 6.2. *Suppose given a block bundle* w/X *and isomorphism* $w_0 \oplus w_1/Y \cong w|Y$, $Y \subset X$. *Then*

$$i_{1!}(i^*(\xi) \cup e(w_0)) = j_1^* j_!(\xi) \quad \text{for } \xi \in T^q(X)$$

where

$$\begin{array}{ccc} Y & \xrightarrow{\;i_1\;} & E(w_1)' \\ \big\downarrow{\scriptstyle i} & & \big\downarrow{\scriptstyle j_1} \\ X & \xrightarrow{\;j\;} & E(w)' \end{array}$$

is the diagram of inclusion, and $E(u)'$ *denotes the pair* $(E(u), \dot{E}(u))$.

Proof. Consider the diagram of inclusions

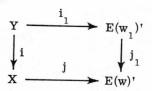

$$\begin{array}{ccccccc}
 & & E(w_0)' & & \xrightarrow{\quad l_1 \quad} & & \\
 & \overset{i_0}{\nearrow} & & \overset{j_0}{\searrow} & & & \\
Y & \xrightarrow{i} & X & \xrightarrow{j} & E(W)' & \xleftarrow{l} & E(W_0 \oplus W_1)' \\
 & \underset{i_1}{\searrow} & & \underset{j_1}{\nearrow} & & & \\
 & & E(w_1)' & & \xrightarrow{\quad l_0 \quad} & &
\end{array}$$

Since $j_1 = l\,l_0$ we have $j_1^*j_!(\xi) = l_0^* l^* j_!(\xi)$. Now apply (II; 2.4) to the square,

$$\begin{array}{ccc}
E(w_0 \oplus w_1)' & \xrightarrow{\quad l \quad} & E(w)' \\
\uparrow{\scriptstyle l_1 i_0} & & \uparrow{\scriptstyle j} \\
Y & \xrightarrow[\quad i \quad]{} & X
\end{array}$$

and get $l_0^* l_{1!} i_{0!} i^*(\xi)$. Again apply (II; 2.4) to the square

$$\begin{array}{ccc}
E(w_0)' & \xrightarrow{\quad l_1 \quad} & E(w_0 \oplus w_1)' \\
\uparrow{\scriptstyle i_0} & & \uparrow{\scriptstyle l_0} \\
Y & \xrightarrow{\quad i_1 \quad} & E(w_1)'
\end{array}$$

and get $i_{1!} i_0^* i_{0!} i^*(\xi)$, which is $i_{1!}(i^*(\xi) \cup e(w_0))$ by (II; 2.6).

Proof of 6.1. Consider the diagram

$$\begin{array}{ccc}
E_\xi \times P_n & \xrightarrow{\quad \Delta_\xi \quad} & (E_\xi \times E_\xi \times S^n)/Z_2 \\
\cap\, i & & \cap\, i' \\
E(E_v \times P_n)' & \xrightarrow{\quad \Delta_v \quad} & E(E_v \times E_v \times S^n)'/Z_2 \\
\uparrow{\scriptstyle c} & & \uparrow{\scriptstyle c'} \\
E(E_u \times P_n)' & \xrightarrow{\quad \Delta_u \quad} & E(E_u \times E_u \times S^n)'/Z_2 \\
\cup\, j' & & \cup\, j' \\
X \times P_n & \xrightarrow{\quad \Delta_x \quad} & (X \times X \times S^n)/Z_2
\end{array}$$

where c, c' are collapsing maps. Let

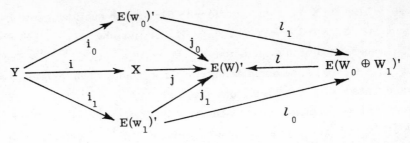

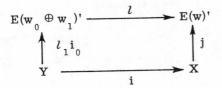

$p : E_\xi \times P_n \to X \times P_n$ and $p' : (E_\xi \times E_\xi \times S^n)/Z_2 \to (X \times X \times S^n)/Z_2$

be the projections. Now apply 6.2 to the top square to get

$$\Delta_v^* i_!' 1 = i_! (\Delta_\xi^*(1) \cup e(v^+)) = i_! (e(v^+)) \ .$$

Now from definitions $i_! p_! = c^* i_!$ and so we have

$$\Delta_v^* i_!' 1 = j_! p_! e(v^+) \ \ldots . \ (1)$$

Now apply 6.2 to the element $p_*'(1)$ and the bottom square to get

$$\Delta_u^* j_!' p_!' (1) = j_! (Sq_0^n(\xi) \cup e(u^+)) \ \ldots . \ (2),$$

since from definitions $\Delta_x^* p_!' (1) = Sq_0^n(\xi)$. Since $j_!' p_!' = c'^* i_!'$ and $\Delta_v c = c' \Delta u$ we have from (1) and (2) that $j_! p_! e(v^+) = j_! (Sq_0^n(\xi) \cup e(u^+))$ and hence the result since $j_!$ is the Thom isomorphism.

Suppose now that u, v are vector bundles. Then u, v have underlying pl structure (see [2]), and it is easy to see that $u^+ = u \otimes l_n$ and $v^+ = v \otimes l_n$ (see for example [1; p. 138]). From 5.1 we now have the following corollary.

Corollary 6.3. If ξ is in u with normal bundle v and u, v admit vector bundle structures then

$$\underline{W}_n(u) Sq_0^n(\xi) = p_! \underline{W}_n(v).$$

Corollary 6.4. For pl bundle u/x with (canonical) Thom class t_u, $Sq_0^n(t_u) = i_! e(u^+)$, where $i : X \times P_n \to X \times E(u)$ is the inclusion.

Proof. Apply 6.2 after replacing $E_\xi \subset E_u \to X$ by $X \subset E_u \xrightarrow{id} E_u$.

7. THE TRANSVERSALITY DEFINITION OF THE EXPANDED SQUARE

The previous section raises several interesting questions, e.g. the relation between u^+ and $u \otimes l_n$ in case u is a pl bundle with Z_2-action. This section clarifies the situation, see e.g. Proposition 7.2

below. This is achieved by a further definition of Sq_0^n as a special case of a general construction which we now describe.

Let $c : \tilde{X} \to X$ be an r-fold covering map. We define a function $P_c : T^q(X) \to T^{rq}(X)$ as follows. Let w_c/X be the vector bundle with fibre at $x \in X$ the vector space with basis set $c^{-1}(x)$. Now let $[\xi] \in T^q(X)$ then for some large N we may assume ξ is in the trivial bundle ε^N with normal bundle v, say. Define ξ_c in $\varepsilon \otimes w_c$ by $E(\xi_c) = \{y \otimes z : y \in E(\xi), \ c(z) = p_\xi(y)\}$. Observe that $\xi_c = c_! c^* \xi$. Now define a (vector) bundle $v_c^\perp/E(\xi_c)$ by taking the fibre over $y \otimes z$ to be the subspace of the fibre of $\varepsilon \otimes w_c$ at $p_\xi(y)$ generated by vectors $a \otimes b$, with $c(b) = p_\xi(y)$ and $b \neq z$. Further define $f : E(v_c^\perp) \to E(\varepsilon \otimes w_c)$ by $f(y \otimes z, \sum a_i \otimes b_i) = y \otimes z + \sum a_i \otimes b_i$. Note that the image of f is a closed subspace of $E(\varepsilon \otimes w_c)$. Now suppose ξ has base K, so $|K| = X$. Over a vertex of K, $E(v_c^\perp)$ falls into r-distinct pieces and under f these pieces intersect in the fibre of $\varepsilon \otimes w_c$ over the vertex. Shift f so the r-pieces and all the intersections are pairwise transverse. This can be done using (II; 4.1). The total intersection is the block of $P_c(\xi_c)$ over the vertex. This is the start of an inductive process which we call 'making f self-transverse'. The resulting self-intersection is the total space of $P_c(\xi)$. The self-intersection should not be confused with the result of intersecting f with itself, which is got by taking two copies of f and intersecting one with the other which would give the cup product $[f] \cup [f]$ where f is regarded as a mock bundle over $E(\varepsilon \otimes W_c)$.

It is not hard to show that the operation is well defined and functorial for bundle maps, and for the trivial r-fold cover P_c is just the r-fold product (see II end of §4).

The description of $P_c(\xi)$ considerably simplifies in case $p_\xi : E_\xi \to X$ is an embedding. In this case one can inductively make $p : E(c_! c^* \xi) \to X$ self transverse. As an example consider the double cover $c : S^n \to P_n$ and $p_\xi : P_{n-1} \to P_n$ the inclusion so that $[\xi] \in T^{-1}(P_n)$ is the generator. Then $P_c(\xi) = 0$ since $S^{n-1} \to P_{n-1} \to P_n$ can be shifted to have empty self intersection.

Theorem 7.1. Let $c : S^n \times X \to P_n \times X$ be the double cover, then $Sq_0^n \xi = P_c(1_{P_n} \times \xi)$.

Proof. Consider ξ in ε^N. Then the quotient map $f : (E_\xi \times (X \times \mathbf{R}^N) \times (X \times \mathbf{R}^N) \times E_\xi) \times S^n \to ((X \times \mathbf{R}^N) \times (X \times \mathbf{R}^N) \times S^n)/\mathbf{Z}_2$ is self transverse with self intersection $(E_\xi \times E_\xi \times S^n)/\mathbf{Z}_2$, which gives $Sq_0^n \xi$ after restriction to $X \times P_n$. But the restriction of the domain of f to $X \times P_n$ is $E(v_c^\perp)$ and its self intersection is $P_c(1_{P_n} \times \xi)$. The result follows from the commutativity of the subdivision with the operation of making self transverse.

We shall use the new definition to prove:

Proposition 7. 2. Suppose that u is a block bundle with $\mathbf{Z}_2$-action. Then $Sq_0^n tu = i_! \underline{W}_n(u)$, where $i : X \times P_n \to E_u \times P_n$ is the inclusion.

Proof. It follows from 7. 1 that $Sq_0^n tu$ can be obtained by making $X \times S^n \to E(u) \times P_n$ self transverse. On the other hand $e(u \otimes l_n)$ is obtained by making $X \times P_n \times S^0 \to X \times P_n \to E(u \otimes l_n)$ self transverse. Both $E(u) \times P_n$ and $E(u \otimes l_n)$ are quotients of $E(u) \times S^n$ under a (free) $\mathbf{Z}_2$-action, by definition. Inductively make $X \times S^n \to E(u) \times S^n$ equivariantly (with respect to the diagonal action on the right) transverse to $X \times S^n$. Suppose the result is $f : X \times S^n \to E(u) \times S^n$. Then $f' : f^{-1}(X \times S^n)/\mathbf{Z}_2 \to X \times P_n$ is the projection of $(u \otimes l_n) = \underline{W}_n(u)$. On the other hand $X \times S^n \xrightarrow{f} E(u) \times S^n \to E(u) \times P_n$ is self transverse and restricting to the intersection gives f' again.

Corollary 7. 3. If u is a block bundle which admits a $\mathbf{Z}_2$-action then

(i) $\underline{W}_n(u)$ is independent of the choice of $\mathbf{Z}_2$-action, and

(ii) if u reduces to a pl bundle then

$$e(u^+) = e(u \otimes l_n) = \underline{W}_n(u).$$

Proof. (i) is immediate and (ii) follows from 6. 4 and 7. 2.

Let X_m be the class of the inclusion $P_{m-1} \subset P_m$ and let $H(m-1, n) \subset P_{m-1} \times P_n \subset P_m \times P_n$ be defined by $\{(x, y) : \sum x_i y_i = 0\}$, see e. g. [1] p. 164.

Corollary 7.4. $Sq_0^n(X_m)$ <u>is the class of the inclusion</u> $H(n, m) \subset P_n \times P_m$.

Proof. X_m is essentially the Thom class $t_{l_{m-1}}$. So $Sq_0^n(X_m) = i_! e(l_{m-1} \otimes l_n)$ but this is $H(m-1, n) \subset P_m \times P_n$ by direct construction, see Theorem 6.6 of [1], p. 164.

Definition 7.5. For any block, vector or pl bundle u define $\underline{W}_n(u)$ by $i_! \underline{W}_n(u) = Sq_0^n(tu)$, where $i : X \times P_n \to E_u \times P_n$.

In view of 7.2 no confusion can arise from this definition.

REFERENCES FOR CHAPTER V

[1] T. Brocker and T. tom Dieck. Kobordismen Theorie. Springer-Verlag lecture notes no. 178.

[2] R. Lashof and M. Rothenberg. Microbundles and smoothing. Topology, 3 (1965), 357-88.

[3] D. Quillen. Elementary proofs of some results of cobordism theory using Steenrod operations. Advances in Math., 7 (1971), 29-56.

[4] C. P. Rourke and B. J. Sanderson. Block bundles II: transversality. Annals of Maths., 87 (1968), 255-77.

[5] T. tom Dieck. Steenrod-operationen in Kobordismen-Theorien. Math. Z., 107 (1968), 380-401.

VI · Sheaves

In this chapter we extend the treatment of coefficients in Chapter III to cover sheaves of abelian groups. We work always with pl cobordism but everything that we say can be extended to an arbitrary theory under the conditions of IV 6. 4. §§4 and 5 in fact extend unconditionally. The general definition of sheaves of coefficients does not have all the best properties one would hope for and we will explain where the difficulties lie at the start of §4.

In §1 we recall the basic properties of stacks and sheaves and in §2 we define the theory of mock bundles with coefficients in a stack. The definition is functorial on the category of all stacks of abelian groups. The main theorem asserts that, if the stack is 'nice', then there is a spectral sequence expressing the relation between simplicial cohomology and cobordism with coefficients in the stack. In §3 cobordism with coefficients in a sheaf is defined by means of a simplicial analogue of the Čech procedure. In §4 we discuss an extension of the methods used in the previous sections and give an example of 'Poincaré duality' between bordism and cobordism with coefficients in the sheaf of local homology of a Z_n-manifold. Finally in §5 we extend the methods further and give examples which suggest the existence of a bordism version of the Zeeman duality spectral sequence [1].

1. STACKS

A stack of abelian groups over a cell complex K is a covariant functor $\tau : \underset{\sim}{K} \to \mathcal{A}b$, where $\mathcal{A}b$ denotes the category of abelian groups and $\underset{\sim}{K}$ denotes the category with objects the cells of K and morphisms the face inclusions. A homomorphism between stacks, τ/K, τ'/K, is a natural transformation of functors $\phi : \tau \to \tau'$. The category of stacks over K will be denoted by $\mathcal{S}/K$ or simply $\mathcal{S}$. If τ is a stack over K and $K' \lhd K$, the <u>subdivision of</u> τ <u>over</u> K' is the stack τ'/K' such that $\tau'(\sigma') = \tau(\sigma)$, $\tau'(\sigma'_1 < \sigma'_2) = \tau(\sigma_1 < \sigma_2)$; $\sigma', \sigma'_1, \sigma'_2 \in K'$, $\sigma, \sigma_1, \sigma_2 \in K$, $\sigma' \subset \sigma$, $\sigma'_1 \subset \sigma_1$, $\sigma'_2 \subset \sigma_2$. If τ/K is a stack, $\tau \times I$ will denote the stack over the cell complex $K \times I$, such that: $(\tau \times I)(\sigma \times I) = (\tau \times I)(\sigma) = \tau(\sigma)$ for every $\sigma \in K$; $(\tau \times I)(\sigma_1 \times I < \sigma_2 \times I) = (\tau \times I)(\sigma_1 < \sigma_2) = \tau(\sigma_1 < \sigma_2)$, $\sigma_1, \sigma_2 \in K$. If τ/K is a stack, and $J \subset K$, the <u>restriction</u> $\tau | J$ is defined to be the stack over J, such that $(\tau|J)(\sigma) = \tau(\sigma)$, $\sigma \in J$; $(\tau|J)(\sigma_1 < \sigma_2) = \tau(\sigma_1 < \sigma_2)$, $\sigma_1, \sigma_2 \in J$.

Let X be a polyhedron and F a presheaf of abelian groups over X. If K is a cell complex, $|K| = X$, F induces a stack, F_K/K, by: $F_K(\sigma) = F(st(\sigma, K))$, $F_K(\sigma_1 < \sigma_2) = F(st(\sigma_1, K) \supset st(\sigma_2, K))$, $\sigma, \sigma_1, \sigma_2 \in K$.

We briefly recall the notion of simplicial cohomology with coefficients in a stack and its relation with Čech cohomology. Let τ/K be a stack over the oriented simplicial complex K. A (-p)-<u>cochain</u>, f^p, with coefficients in τ, is a map which assigns to each p-simplex $\sigma^p \in K$, an element of $\tau(\sigma^p)$; (-p)-cochains form an abelian group by coordinate addition, $C^{-p}(K, \tau)$. There is a coboundary homomorphism: $\delta^{-p} : C^{-p}(K, \tau) \to C^{-p-1}(K, \tau)$, given by $\delta^p f^p(\sigma^{p+1}) = \sum_{\sigma \leq \sigma^{p+1}} [\sigma : \sigma^{p+1}] \tau(\sigma < \sigma^{p+1}) f^p(\sigma)$. $\{C^p(K, \tau), \delta^p\}$ is a chain complex and its cohomology is by definition $H^p(K, \tau)$.

If F/X is a sheaf and F_K the induced stack over K, let $\mathfrak{U}_K$ denote the covering of X formed by the stars of the vertices of K. The nerve of $\mathfrak{U}_K$ is known to be equal to K and this fact induces a canonical identification $\{C^p(K, F_K), \delta^p\} \overset{\cong}{\to} \{\check{C}^p(\mathfrak{U}_K, F), \check{\delta}\}$ of chain complexes; $\check{}$ stands for 'Čech'. Therefore, because the coverings by stars are a cofinal system in the system of all coverings of X, we get

$$\lim_{\to K} H^p(K; F_K) = \lim_{\to \mathfrak{U}} \check{H}^p(\mathfrak{U}, F) = \check{H}^p(X; F).$$

2. COBORDISM WITH COEFFICIENTS IN A STACK

Throughout this section we shall be using the same terminology as in Chapter III below 3.5. Thus a G-<u>cycle</u> or G-<u>manifold</u> has singularities in codimension 1 at most, while a G-<u>bordism</u> is allowed to have singularities up to codimension 2.

Let τ be a stack of abelian groups over an oriented cell complex K. A (τ, q)-cocycle [cobordism] ξ^q/K consists of a projection $p_\xi : E \to |K|$ such that

(a) for each $\sigma \in K$, $p_\xi^{-1}(\sigma)$ is the interior of a $(\tau(\sigma), q+\dim \sigma)$-manifold [bordism], $\overline{\xi(\sigma)}$

(b) for each $\sigma \in K$, $\overline{\xi(\sigma)} = \bigcup_{\sigma_i < \sigma} [\sigma_i : \sigma]\tau(\sigma_i < \sigma)p_\xi^{-1}(\sigma_i)$, where $\tau(\sigma_i < \sigma)p_\xi^{-1}(\sigma_i)$ is the image of the $\tau(\sigma_i)$-manifold [bordism] under the relabelling morphism $\tau(\sigma_i < \sigma)$; $[\sigma_i : \sigma]$ is the incidence number and its effect is a change of orientation iff $[\sigma_i : \sigma] = -1$.

The manifold [bordism] $\xi(\sigma) = \overline{\xi(\sigma)} - \partial\overline{\xi(\sigma)}$ is called the block over σ; (τ, q)-cocycles [cobordisms] ξ, η over K are isomorphic, written $\xi \cong \eta$, if there is a (pl) homeomorphism $h : E_\xi \to E_\eta$ such that $h|\xi(\sigma)$ is an isomorphism of $\tau(\sigma)$-manifolds [bordisms] between $\xi(\sigma)$ and $\eta(\sigma)$. Suppose given ξ^q/K and $L \subset K$, then the restriction $\xi|L$ is defined in the obvious way and it is a $(\tau|L, q)$-cocycle [cobordism] over L. A (τ, q)-cocycle [cobordism] over (K, L) $(L \subset K)$ is a (τ, q)-cocycle [cobordism] over K, which is empty over L; $-\xi$ is the cocycle [cobordism] obtained from ξ by reversing the orientation in each block. If ξ_0 and ξ_1 are (τ, q)-cocycles over (K, L), then ξ_0 is cobordant to ξ_1 if there exists a $(\tau \times I, q)$-cobordism $\eta/(K\times I, L \times I)$ such that $\eta|K \times \{i\} = (-1)^i\xi_i$ $(i = 0, 1)$. Cobordism is an equivalence relation and we define $\Omega^q(K, L; \tau)$ to be the set of cobordism classes of (τ, q)-cocycles over (K, L); $\Omega^q(K, L, \tau)$ is an abelian group under the operation of disjoint union; we call it the q-th oriented (pl) cobordism group of (K, L) with coefficients in τ. If $\tau' \lhd \tau$ then there exists an 'amalgamation' homomorphism $am : \Omega^q(K', \tau') \to \Omega^q(K, \tau)$ like in the case of ordinary mock bundles (see Chapter II). In the proof of the subdivision theorem for cobordism without coefficients, Chapter II 2.1, all the geometric constructions are carried out cellwise. Therefore the proof readily adapts to the present case and we deduce

Proposition 2.1. If $\tau' \lhd \tau$ then $am : \Omega^q(K', \tau') \to \Omega^q(K, \tau)$ is an isomorphism of abelian groups.

Let $\phi : \tau_1 \to \tau_2$ be a homomorphism of stacks over K. There is an induced homomorphism $\phi_* : \Omega^*(K; \tau_1) \to \Omega^*(K, \tau_2)$ defined blockwise by relabelling; more precisely, if ξ/K is a τ_1-cocycle [cobordism] and $\sigma \in K$, we relabel the block $\xi(\sigma)$ by means of $\phi(\sigma) : \tau_1(\sigma) \to \tau_2(\sigma)$ (like in III). If $\xi'(\sigma)$ is the resulting polyhedron, then the union $\xi' = \bigcup_{\sigma \in K} \xi'(\sigma)$ is a τ_2-cocycle [cobordism]. All compatibility conditions are ensured by $\phi : \tau_1 \to \tau_2$ being a stack-homomorphism. Therefore we have the following:

Proposition 2.2. There is a functor $\Omega^*(K, -) : \mathcal{S} \to \mathcal{A}b_*$ which assigns to each $\tau \in \mathcal{S}$ the abelian group $\Omega^*(K; \tau)$ and to each morphism $\phi \in \mathcal{S}$ the (graded) abelian-group homomorphism ϕ_*.

A linked stack of resolutions over a cell complex K consists of a covariant functor $\rho : \underset{\sim}{K} \to \mathcal{C}$ (see III §3 for the definition of $\mathcal{C}$). A linked stack of resolutions ρ is said to be p-canonical if $\rho(\sigma)$ is a p-canonical linked resolution for each $\sigma \in K$ (in the sense of III §3).

If τ/K is a stack of abelian groups and ρ/K is a linked stack of resolutions such that, for each $\sigma \in K$, $\rho(\sigma)$ is a resolution of $\tau(\sigma)$, then we say that τ is represented by ρ. We denote $\mathcal{S}'/K$ the full subcategory of $\mathcal{S}/K$ consisting of all stacks which are representable by p-canonical stacks for each $p = 1, 2, 3$. The objects of $\mathcal{S}'$ will also be called nice stacks.

If τ/K is a stack of abelian groups, there is an induced graded stack on K, written $\{\Omega^q_\tau/K\}$ and given by

$$\Omega^q_\tau(\sigma) = \Omega^q(\text{point}; \tau(\sigma))$$

$$\Omega^q_\tau(\sigma_1 < \sigma_2) = \tau(\sigma_1 < \sigma_2)_* .$$

We aim to prove the following

Theorem 2.3. If $\tau \in \mathcal{S}'$, there is a spectral sequence, $E(\tau)$, running

$$E^2_{p,q} = H^p(K; \Omega^q_\tau) \Rightarrow \Omega^*(K; \tau).$$

Moreover the spectral sequence is natural on $\mathcal{S}'$.

In order to prove the theorem we need some definitions and lemma

Let ρ be a linked stack over K. A (ρ, q)-mock bundle ξ^q over

(K, L) consists of a projection $p_\xi : E_\xi \to |K|$ such that

(a) for each $\sigma \in K$, $p_\xi^{-1}(\sigma)$ is the interior of a $(\rho(\sigma), q+\dim\sigma)$-manifold, $\overline{\xi(\sigma)}$;

(b) for each $\sigma \in K$, $\overline{\xi(\sigma)} = \bigcup_{\sigma_i < \sigma} [\sigma_i : \sigma]\rho(\sigma_i < \sigma)p_\xi^{-1}(\sigma_i)$ (with the usual meaning of the notations);

(c) $p^{-1}(L) = \emptyset$.

Two (ρ, q)-mock bundles $\xi_0, \xi_1/(K, L)$ are <u>cobordant</u> if there exists a (ρ, q)-mock bundle $\eta/(K \times I, L \times I)$ such that $\eta | K_i = (-1)^i \xi_i$ $i = 0, 1$. The <u>(q-th)-cobordism group</u> of (K, L) with coefficients ρ, written $\Omega^q(-; \rho)$, is constructed from (ρ, q)-mock bundles in the usual fashion.

Thus the main difference between the theory of (τ, q)-cocycles and the theory of (ρ, q)-mock bundles is that the former allows the cobordisms to have deeper singularities than the cocycles while the latter is the natural extension to the case of local coefficients of an ordinary mock-bundle theory.

Lemma 2.4. <u>There exists a coboundary homomorphism</u> $\delta^q : \Omega^q(L; \rho) \to \Omega^{q-1}(K, L; \rho)$ <u>and a long exact sequence</u>

$$\dots \to \Omega^q(K, L; \rho) \xrightarrow{g} \Omega^q(K; \rho) \xrightarrow{f} \Omega^q(L; \rho) \xrightarrow{\delta^q} \dots$$

<u>where</u> g <u>is induced by</u> $(K, L) \to (K, \emptyset)$ <u>and</u> f <u>is 'restriction to</u> L'.

Proof. Definition of δ^q. It can be roughly described as 'pull back onto the boundary of a regular neighbourhood of L in K'. Suppose L full in K. Let $J(\frac{1}{2})$ be the $\frac{1}{2}$-nhd. of L in K and $\dot{J}$ its frontier; $\pi : J \to L$ the pseudo-radial retraction, $\dot{\pi} = \pi | \dot{J}$. If ξ is a (ρ, q)-mock bundle over L, form the pull back $\dot{\pi}^*(\xi)$, which is a q-mock bundle over $\dot{J}$ and a (q - 1)-mock bundle over (K, L). If $\xi^*(\sigma)$ is the block of $\dot{\pi}^*(\xi)$ over σ, then $\xi^*(\sigma)$ comes from the block over $\dot{\pi}(\dot{J}(\frac{1}{2}) \cap \sigma)$, denoted $\dot{\pi}(\sigma)$; therefore it has a structure of $\rho(\dot{\pi}(\sigma))$-manifold and it is made into a $\rho(\sigma)$-manifold by means of the stack homomorphism $(\rho(\dot{\pi}(\sigma) < \sigma)$. So the resulting object $\dot{\pi}^*(\xi)/K$ is a $(\rho, q-1)$-mock bundle which is empty over L. The assignment $[\xi] \to [\dot{\pi}^*(\xi)/K]$ gives a well defined morphism $\delta^q : \Omega^q(L) \to \Omega^{q-1}(K, L)$. If L is not full, one first subdivides barycentrically once and then amalgamates.

Exactness is proved geometrically by mock-bundle arguments, which are all contained in Chapter II. The only remark to make is the following.

Suppose $K = \sigma$ and ξ is a (ρ, q)-mock bundle defined on $J = \dot{\sigma} - \sigma_1$, σ_1 face of σ. Then ρ gives morphisms going from the resolutions attached to the simplexes of J to the resolutions attached to σ. Therefore, if ξ is extended over σ by the pull back construction, the resulting block $\xi(\sigma)$ has a natural structure of $\rho(\sigma)$-manifold.

We remark that the coboundary homomorphism δ^q can be defined also for the theory $\Omega^*(K, L; \tau)$ $(\tau \in \mathcal{S}/K)$ exactly in the same way as in the above proof. Therefore there is, for the theory $\Omega^*(K, L; \tau)$, a long sequence analogous to that of Lemma 2.4. However, although it is immediately checked that the sequence has order two, there is no reason to suppose that it is exact. The argument which is used to prove Lemma 2.4 fails because in $\Omega^*(K, L; \tau)$ two cobordisms having the same ends cannot be glued together to give a cocycle.

Given a linked stack of resolutions ρ/K there is an associated graded stack $\Omega^q_\rho \in \mathcal{S}$ defined like Ω^q_τ above $(\tau \in \mathcal{S})$. We then have the following:

Lemma 2.5. <u>If ρ is a linked stack of resolutions there exists a spectral sequence $E(\rho)$ running</u>

$$H^p(K; \Omega^q_\rho) \Rightarrow \Omega^*(K; \rho).$$

Proof. By Lemma 2.4, for each $p = 0, 1, \ldots$ we have the exact 'p-sequence'

$$\ldots \Omega^{-p-q+1}(K_p, K_{p-1}; \rho) \xrightarrow{g} \Omega^{-p-q+1}(K_p; \rho) \xrightarrow{f} \Omega^{-p-q+1}(K_{p-1}; \rho)$$
$$\xrightarrow{\delta} \Omega^{-p-q}(K_p, K_{p-1}; \rho) \rightarrow \ldots$$

where K_p is the p-skeleton of K. The theory of exact couples yields a spectral sequence $E(\rho)$ in which the chain complex $\{E^1_{p,q};\ d^1_{p,q} :$ q fixed$\}$ is isomorphic to $\{C^p(K; \Omega^q_\rho), \delta\}$ defined in Section 1. More precisely an isomorphism $h : E^1_{p,q} \rightarrow C^p(K; \Omega^q_\rho)$ is given as follows. $E^1_{p,q} = \Omega^{-p-q}(K_p, K_{p-1}; \rho)$. Let $[\xi] \in \Omega^{-p-q}(K_p, K_{p-1}; \rho)$. Then ξ

has a block $\xi(\sigma)$ for each p-simplex $\sigma \in K_p$ and $\xi(\sigma)$ is a closed $(\rho(\sigma), -q)$-manifold, because ξ is empty over K_{p-1}. Therefore $\xi(\sigma)$ determines an element $[\xi(\sigma)]_\rho \in \Omega^q_\rho(\sigma)$. We associate to $[\xi]$ a p-cochain $f^p_{[\xi]}$ with coefficients Ω^q_ρ by setting

$$f^p_{[\xi]}(\sigma) = [\xi(\sigma)]_\rho \quad \text{for each p-simplex } \sigma \in K.$$

It is readily checked that the correspondence $[\xi] \to f^p_{[\xi]}$ defines the isomorphism h. Moreover the required convergence conditions hold and therefore E^∞ is the bigraded module associated to the filtration of $\Omega^*(K; \rho)$ defined by

$$F^s\Omega^*(K; \rho) = \mathrm{Ker}[\Omega^*(K; \rho) \to \Omega^*(K_{p-1}; \rho)].$$

The lemma follows.

Proof of Theorem 2.3. For each $i = 1, 2, 3$, let ρ_i be an i-canonical linked stack representing a given $\tau \in \mathcal{S}'$.

There is a commutative diagram of degree-zero homomorphisms:

$$\begin{array}{ccc} \Omega^*(K; \rho_1) & \xrightarrow{t_{1,2}} & \Omega^*(K; \rho_2) \\ & {\scriptstyle t_{1,3}} \searrow \quad \swarrow {\scriptstyle t_{2,3}} & \\ & \Omega^*(K; \rho_3) & \end{array}$$

in which $t_{i,j}$ is the 'relabelling' map on cocycles. It is easy to see that $t_{i,j}$ commutes with the coboundary operation and therefore there is an induced commutative diagram of spectral-sequence homomorphisms

$$\begin{array}{ccc} E(\rho_1) & \xrightarrow{t^*_{1,2}} & E(\rho_2) \\ & {\scriptstyle t^*_{1,3}} \searrow \quad \swarrow {\scriptstyle t^*_{2,3}} & \\ & E(\rho_3) & \end{array}$$

On the E^2-term $t^*_{i,j}$ is the homomorphism

$$H^p(K; \Omega^q_{\rho_i}) \xrightarrow{t^*_{i,j}} H^p(K; \Omega^q_{\rho_j})$$

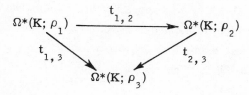

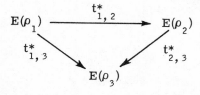

induced by the change of coefficients $\Omega^q_{\rho_i} \xrightarrow{\text{relabel}} \Omega^q_{\rho_j}$ which we know to be an isomorphism from the theory of coefficients in the constant case.

Then, by the usual spectral-sequence argument, $t^*_{i,j}$ is an isomorphism for all $(i, j) : 1 \leq i < j \leq 3$. Now there is a homomorphism $\alpha : \Omega^*(K; \tau) \to \Omega^*(K; \rho_3)$ given by the relabelling map on the cocycles. In order to prove the theorem we only need to show that α is an isomorphism.

1. *α is an epimorphism.* There is a commutative diagram

$$\begin{array}{ccc} \Omega^q(K; \rho_1) & \xrightarrow{t_{1,3}} & \Omega^q(K; \rho_3) \\ & {\scriptstyle\theta}\searrow \quad \nearrow{\scriptstyle\alpha} & \\ & \Omega^q(K; \tau) & \end{array}$$

where θ is also a relabelling map. Therefore, since $t_{1,3}$ is epi, so is α.

2. *α is a monomorphism.* Let $[\xi] \in \operatorname{Ker} \alpha$. It follows from the definitions that ξ/K is also a (ρ_2, q)-mock bundle. Therefore it determines a class $[\xi]_{\rho_2}$ which is mapped to zero by $t_{2,3}$. Because $t_{2,3}$ is a monomorphism there exists a (ρ_2, q)-cobordism $\eta : \xi \sim \emptyset$. But η is also a (τ, q)-cobordism. Therefore α is mono.

The theorem follows.

Corollary 2.6. *If τ/K is a constant stack (i.e. $\tau(\sigma) = G$ for each $\sigma \in K$), then $\Omega^*(K; \tau)$ coincides with $\Omega^*(K; G)$ as defined in II 3.*

Proof. Since $\Omega^*(K; G)$ is a cohomology theory, there is a spectral sequence $E(G)$ running

$$H^p(K, \Omega^*(\text{point}, G)) \Rightarrow \Omega^*(K; G).$$

There is a natural transformation of cohomology theories

$$t : \Omega^*(K; G) \to \Omega^*(K; \tau)$$

given by relabelling. The induced homomorphism of spectral sequences

$E(G) \to E(\tau)$ is an isomorphism on the E^2-term because $\Omega^*_\tau = \Omega^*(\text{point}, G)$.

Therefore the corollary follows from the 'mapping theorem between spectral sequences'.

Corollary 2.7. If $\tau \in \mathcal{S}'$ and we take the theory $\Omega^*(K)$ to be $H^*(K, Z)$ (= simplicial cohomology with Z-coefficients), then $\Omega^*(K, \tau)$ coincides with the usual definition of simplicial cohomology with coefficients in τ (see Section 1).

Proof. $H^*(K; Z)$ is the mock-bundle theory whose blocks are oriented pseudomanifolds. Since a G-pseudomanifold of dimension greater than zero is bordant to $\emptyset$ for every group G, we see that the E^1-term of the spectral sequence $E(\tau)$ reduces to the cochain complex $C^*(K, \tau)$ considered in Section 1 and the spectral sequence collapses.

3. COBORDISM WITH COEFFICIENTS IN A (PRE)-SHEAF

We are now ready to give a notion of cobordism with coefficients in a presheaf, using an analogue of the Čech procedure. Let X be a polyhedron and F/X a presheaf of abelian groups. If K is a triangulation of X, then, by the previous section, we have a graded group $\{\Omega^q(K, F_K)\}$, where F_K is the induced stack on K. Suppose $K' \lhd K$. We define a homomorphism $\alpha_{K,K'} : \Omega^q(K, F_K) \to \Omega^q(K', F_{K'})$ as follows: let ξ^q/K be an (F_K, q)-mock bundle. Subdivide ξ over K' and get ξ''/K' such that $\xi''(\sigma')$ is an $F_K(\sigma)$-manifold for $\sigma' \subset \sigma$. The inclusion $st(\sigma', K') \subset st(\sigma, K)$ gives a restriction homomorphism $F_{\sigma,\sigma'} : F_K(\sigma) \to F_{K'}(\sigma')$ and we make $\xi''(\sigma')$ into an $F_{K'}(\sigma')$-manifold, $\xi'(\sigma')$, by applying this homomorphism, i.e. $\xi'(\sigma') = F_{\sigma,\sigma'}(\xi''(\sigma'))$. When all the blocks of ξ'' have been relabelled by means of the restriction homomorphisms, one takes care of the orientations in the blocks, so that the incidence numbers are preserved. The functoriality of the presheaf F ensures that the final object is an $(F_{K'}, q)$-mock bundle ξ'/K', called an F-subdivision of ξ over K'. Two F-subdivisions ξ', $\bar{\xi}'$ of ξ over K' are cobordant by the same construction applied to $K \times I$ and $K' \times I$ modulo the ends. Therefore we have a well defined homomorphism

$$\alpha_{K,K'} : \Omega^q(K, F_K) \to \Omega^q(K', F_{K'})$$
$$\alpha_{K,K'} : [\xi] \to [\xi'].$$

The collection of groups and homomorphisms $\{\Omega^q(K, F_K), \alpha_{K, K'}\}$, indexed by the directed set of all triangulations of X, is a direct system and we define the q-th (pl) cobordism group of X with coefficients in F to be the graded group:

$$\Omega^q(X, F) = \lim_{\to K} \{\Omega^q(K, F_K), \alpha_{K,K'}\}.$$

We now recall that a sheaf F/X is locally constant if there is an open covering $\mathfrak{A} = \{U\}$ of X, such that, if $U \in \mathfrak{A}$ and $x \in U$ then $F(U) = \lim_{\to} \{F(V)\}$ where V varies over the open neighbourhoods of x. It then follows that, if K is a sufficiently small triangulation of X, i.e. the associated star-covering is a refinement of $\mathfrak{A}$, the cohomology of K with coefficients in the stack F_K coincides with the Čech-cohomology $\check{H}^*(X; F_K)$ and the cobordism of K with coefficients in F_K coincides with $\Omega^*(X; F)$ by 2.1. We call such an F_K/K a limit stack for F/X. We say that a locally constant sheaf F/X is nice if it has a limit stack which is nice in the sense of Section 2.

As in the case of stacks, to each sheaf F/X there is associated a graded sheaf $\{\Omega^q_F/X\}$ defined by $\Omega^q_F(U) = \Omega^q(\text{point}; F(U))$, $\Omega^q_F(U \supset V) = F(U \supset V)_*$; U, V open sets of X. If F/X is locally constant, then Ω^q_F/X is also locally constant.

We have the following analogue of Theorem 2.3.

Theorem 3.1. On the category of nice sheaves there is a natural spectral sequence running

$$\check{H}^p(X; \Omega^q_F) \Rightarrow \Omega^*(X; F).$$

The proof is the same as that of 2.3, using limit stacks.

A direct consequence of the 'mapping theorem between spectral sequences' is the following comparison theorem

Proposition 3.2. Let F^*/X be a nice sheaf. If $h : T^*(-) \to S^*(-)$

is a natural transformation of cohomology theories, inducing an isomorphism of the corresponding graded sheaves

$$h_F : T^*_F \to S^*_F$$

then h induces an isomorphism in local coefficients:

$$h^*(X) : T^*(X; F) \to S^*(X; F) .$$

4. QUASI-LINKED STACKS

We have given a definition of 'pl cobordism with coefficients in a sheaf' which works in the category of all sheaves, is functorial on this category and extends the case of coefficients in an abelian group. Unfortunately, we can be sure that the definition enjoys good properties (e.g. spectral sequence) only when some special types of sheaves are involved. Let us try to describe the source of this difficulty. A mock bundle and a stack, τ, over a simplicial complex K, have a main feature in common: they are both 'local' objects in the sense that they are both functors defined on the category $\underset{\sim}{K}$. A mock bundle takes its values in the category of pl manifolds and inclusions in the boundary, while a stack of abelian groups ranges in the category of abelian groups and homomorphisms. Thus in order to have a notion of mock bundle with τ coefficients, which has good properties, what is needed is the following:

1. a notion of $\tau(\sigma)$-manifold ($\sigma \in K$) for which the corresponding mock bundle theory is 'cobordism with $\tau(\sigma)$-coefficients'.

2. a recipe for associating a $\tau(\sigma_2)$-manifold to a $\tau(\sigma_1)$-manifold in a way which is functorial on $\underset{\sim}{K}$.

Therefore we see that the difficulties arising from our definition can be traced to the lack of a bordism theory of G-cycles which is functorial (with respect to G) on the cycles, rather than only on the bordism classes. Now in many instances it happens that conditions 1 and 2 can be fulfilled and since some of the cases look rather interesting, we will make a detailed discussion of them.

Thus, in the remainder of the chapter, we abandon the point of view of setting up a theory of local coefficients in generality. Instead we discuss, along the lines of the above remarks, some generalisations of our method and give examples to show how local coefficients may at times reveal relationships between local and global properties of a space. We continue to work only with pl cobordism, but everything that we say remains true for an arbitrary geometric cohomology theory.

Let ρ, ρ' be linked resolutions and $f : \rho \to \rho'$ a chain map. f is said to be quasi-linked if the following conditions are satisfied:

(a) For each $b^p \in B^p$, $p = 0, 1, 2, 3$ either $f(b^p) = 0$ or one of $\pm f(b^p)$ is in B'^p.

(b) For each link $L(b^p, \rho)$ let $fL(b^p, \rho)$ be obtained from $L(b^p, \rho)$ by the following process. Let $V \subset L(b^p, \rho)$ be a stratum labelled by b^j: if $f(b^j) = 0$, remove V from $L(b^p, \rho)$; if $f(b^j) \in B'^j$, relabel V by $f(b^j)$; if $-f(b^j) \in B'^j$, reverse the orientation of V and then label $-V$ by $-f(b^j)$. Then we require $fL(b^p, \rho) = \delta L(\delta f b^p, \rho)$ for one of $\delta = \pm$.

To each (ρ, n)-manifold M there is associated a (ρ', n)-manifold $f(M)$ constructed on strata in the same way as $fL(b^1, \rho)$. There is a category Q, whose objects are short linked resolutions and whose morphisms are quasi-linked maps. A quasi-linked stack of resolutions over a ball complex K is a covariant functor $S = S_K : \underset{\sim}{K} \to Q$. If K is oriented, an (S, q)-mock bundle ξ^q over K consists of a projection $p_\xi : E_\xi \to |K|$ such that, for each $\sigma \in K$, $p_\xi^{-1}(\sigma)$ is the interior of an $(S(\sigma), q + \dim \sigma)$-manifold $\overline{\xi(\sigma)}$, with $\partial\overline{\xi(\sigma)} = \underset{\sigma_i < \sigma}{\smile} [\sigma_i : \sigma] S(\sigma_i < \sigma) p_\xi^{-1}(\sigma_i)$.

The cobordism theory of (S, q)-mock bundles is developed exactly in the same way as the theory of cobordism with coefficients in a linked stack of resolutions (see §2). In particular we have

(a) an abelian group $\Omega^q(K, L; S)$, the q-th cobordism group of K, L ($L \subset K$) with coefficients in S, depending only on S, $|K|$, $|L|$.

(b) a spectral sequence running

$$H^p(|K|, \Omega_S^q) \Rightarrow \Omega^*(K, S)$$

where Ω_S^q is the stack of abelian groups over K defined by

$$\Omega_S^q(\sigma) = \Omega^q(\text{point}, S(\sigma))$$

and $\Omega^q_s(\sigma_1 < \sigma_2) : [M] \to [S(\sigma_1 < \sigma_2)M]$.

As an illustration of the general setting described above, we discuss some examples associated with Poincaré duality.

Let (P, SP) be a Z_n-manifold of dimension m (see III 1.1(2)), triangulated by (K, SK)(SK $\subset$ K) and consider the stack $\mathcal{L} = \mathcal{L}(K)$ of local m-homology on K.

If $n = 2$, then a presentation of $\mathcal{L}$ is the following:

(a) $\mathcal{L}(\sigma) =$ free abelian group, $F(x_\sigma)$, on one element x_σ

(b) if σ', $\sigma \in K - SK$ or σ', $\sigma \in SK$ and $\sigma' < \sigma$, then $\mathcal{L}(\sigma' < \sigma)$ is the isomorphism mapping $x_{\sigma'}$ to x_{σ};

(c) if $\sigma' \in SK$, $\sigma \in K - SK$, $\sigma' < \sigma$, then $\mathcal{L}(\sigma' < \sigma)$ is one of the isomorphisms $x_{\sigma'} \to \pm x_{\sigma'}$, depending on which 'sheet' σ belongs to. We call 'positive' (resp 'negative') the sheet corresponding to the '+' sign (resp. '-' sign).

For a Z_3-manifold, $\mathcal{L}$ can be presented as follows. If $\sigma \in K - SK$, then $\mathcal{L}(\sigma) = F(x_\sigma)$; if $\sigma \in SK$, then $\mathcal{L}(\sigma) = F(x^\sigma_1, x^\sigma_2)$. The morphisms are so defined.

(a) if σ', $\sigma \in K - SK$ or σ', $\sigma \in K$ and $\sigma' < \sigma$ then $\mathcal{L}(\sigma' < \sigma)$ is the canonical isomorphism;

(b) if $\sigma' \in SK$, and $\sigma \in K - SK$; $\sigma' < \sigma$, then $\mathcal{L}(\sigma' < \sigma)$ is one of the following isomorphisms, depending on which sheet σ belongs to:

$$(1) \begin{cases} x^{\sigma'}_1 \to x_\sigma \\ x^{\sigma'}_2 \to x_\sigma \end{cases} \qquad (2) \begin{cases} x^{\sigma'}_1 \to -x_\sigma \\ x^{\sigma'}_2 \to 0 \end{cases} \qquad (3) \begin{cases} x^{\sigma'}_1 \to 0 \\ x^{\sigma'}_2 \to -x_\sigma \end{cases}$$

The sheet for which the assignment (i) holds (i = 1, 2, 3) will be referred to as 'sheet (i)'.

Associated to $\mathcal{L}$, there is the following quasi-linked stack, Λ, on the Z_3-manifold K: if $\sigma \in K - SK$, then $\Lambda(\sigma)$ is $F(x_\tau) \overset{id}{\rightarrowtail\!\!\!\twoheadrightarrow} F(x_\sigma)$; if $\sigma \in SK$, then $\Lambda(\sigma)$ is $F(x^\sigma_1, x^\sigma_2) \overset{id}{\rightarrowtail\!\!\!\twoheadrightarrow} F(x^\sigma_1, x^\sigma_2)$; $\Lambda(\sigma' < \sigma) = \mathcal{L}(\sigma' < \sigma)$. We write $\Omega^q(P, \hat{\mathcal{L}}) = \varinjlim \Omega^q(K, \Lambda)$. Here $\hat{\mathcal{L}}$ stands for 'local-homology sheaf'.

With the above notations we have the following

Proposition 4.1. <u>There is a duality isomorphism:</u>

$$\psi : \Omega^q(P, \hat{\mathcal{L}}) \to \Omega_{m+q}(P).$$

Proof. $n = 2$. In this case SP is an orientation-type singularity. Let $[\xi] \in \Omega^q(K, \Lambda)$. The total space $E(\xi)$ is a manifold (see Chapter II). Moreover $E(\xi)$ is oriented, because it is oriented over both $K - SK$ and SK and the action of $\mathcal{L}$ makes the orientations of the blocks over the positive sheet compatible with those of the blocks over the negative sheet. Therefore we regard $E(\xi)$ as an oriented bordism class and define a 'glueing map' ψ like in Chapter II. The proof that ψ is an isomorphism reduces essentially to the proof of Poincaré duality given in II and we omit it.

$n \geq 3$. Again, for the sake of simplicity, we only discuss the case $n = 3$. The general case is dealt with using the same arguments. Let $[\xi] \subset \Omega^q(K, \Lambda)$. The block of ξ over a simplex $\sigma' \in SK$ is a pl manifold, each component of which is labelled by either $x_1^{\sigma'}$ or $x_2^{\sigma'}$ and at most two non-empty blocks merge into it, so that no singularities are created in the glueing process (see Fig. 18). Moreover, as in the above case, the signs in the stack-homomorphisms ensure that the orientations of the blocks are compatible in passing from one sheet to another across SK. Hence the total space $E(\xi)$ is an oriented pl manifold and the operation of 'glueing up and disregarding the labels' gives the required homomorphism ψ.

We now prove that ψ is an epimorphism. Let $f : W^{m+q} \to K$ be a simplicial map representing an oriented bordism class of P. Then Cohen's generalized transversality theorem (see IV 1.1(2)), together with the subdivision theorem, tells us that there exists a cell decomposition (L, SL) of (P, SP) and $\bar{f} \simeq f$, such that

(a) $\bar{f} | \bar{f}^{-1}(L - \text{Int } L_0)$ is the projection of a pl oriented q-mock bundle, where L_0 is a subcomplex of L triangulating a product neighbourhood of SP in P, i.e. $|L_0| = SP \times \text{cone (3 pts)}$. In order to avoid technical details, we assume that L_0 is the product cone-complex $SL \times \text{cone (3 pts)}$ and write L_0^i for $L_0 \cap \text{Sheet (i)}$ $(i = 1, 2, 3)$.

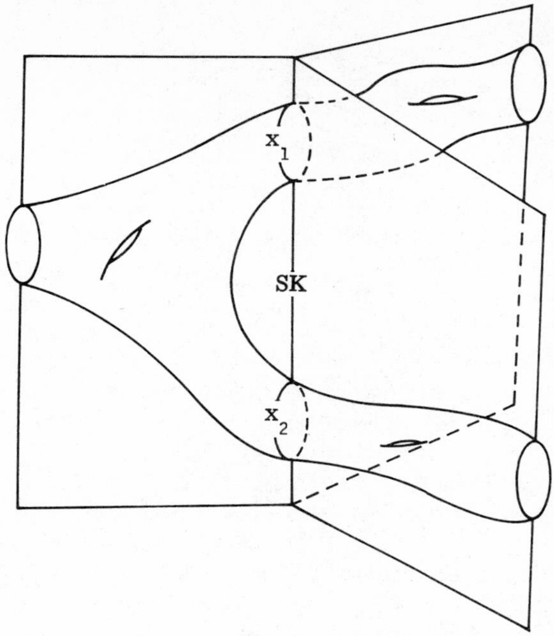

Fig. 18

(b) $\bar{f}^{-1}(L_0)$ is an oriented manifold with boundary $\bar{f}^{-1}(\partial L_0)$ and $\bar{f}|\bar{f}^{-1}(L_0)$ is the projection of a q-mock bundle over the cone complex $L_0 = \{\sigma \times \text{cone (3 pts)} : \sigma \in SL\}$.

Now we take boundary collars $\bar{f}^{-1}(\partial L_0^j) \times I \subset \bar{f}^{-1}(L_0)$ $(j = 2, 3)$ and 'stretch' them along the cone-lines of L_0 until $\bar{f}$ is replaced by a map $f' : W^{m+q} \to |L|$ such that, with obvious meaning of the notations:

(a) $f' \simeq \bar{f}$ mod $L - \text{Int } L_0$

(b) $f'^{-1}(SL) = \bar{f}^{-1}(\partial L_0^j) \times 1 \subset \bar{f}^{-1}(\partial L_0^j \times I)$ $(j = 2, 3)$. (See Fig. 19.)

Then, for $\sigma' \in SL$ and $f'^{-1}(\sigma') \neq \emptyset$, we label the block $f'^{-1}(\sigma')$ by $x_1^{\sigma'}$ if the two non-empty blocks merging into it lie over the sheets (1) and (2) respectively; otherwise we label $f'^{-1}(\sigma')$ by $x_2^{\sigma'}$. For each $\sigma \in L - SL$ we give $f'^{-1}(\sigma)$ the label x_σ. In this way $f' : W^{m+q} \to |L|$ gives rise to a q-mock bundle associated to Λ_L. This shows that ψ

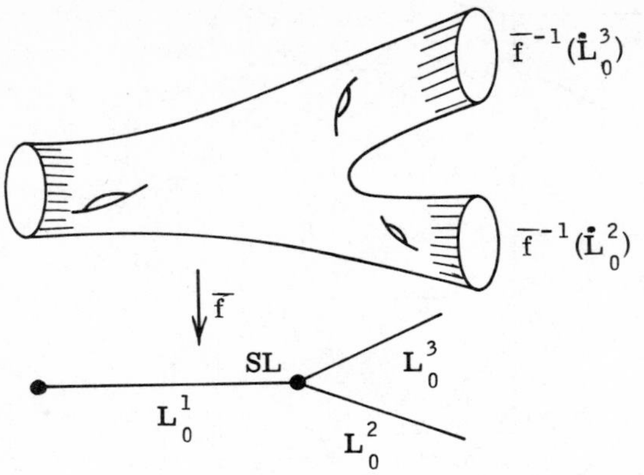

Fig. 19

is an epimorphism. The injectivity of ψ follows from the same arguments applied to $P \times I$.

The proof of the proposition is complete.

Another example of a polyhedron, whose local homology gives rise to a quasi-linked stack in a natural way, is provided by any unoriented pl manifold M^m. We leave the reader to make the obvious definition of $\Omega^q(M, \hat{\mathcal{L}})$ in this case, while we establish the following fact about M, which is an easy consequence of Proposition 4.1.

Corollary 4.2. Suppose that the orientation cover $\mathcal{O}(M^m)$ of M is isomorphic to $f^*(\eta)$, where η is the non-trivial double covering of the circle S^1 and $f : M \to S^1$ is a map. Then there is a duality isomorphism,

$$\psi : \Omega^q(M, \hat{\mathcal{L}}) \to \Omega_{m+q}(M) .$$

Proof. We regard M as a Z_2-manifold in the following way. First we make f transverse to a point $x_0 \in S^1$, then take $SM = f^{-1}(x_0)$ and give orientations to $M - SM$ and SM. The fact that SM is orientable follows immediately from the equation.

$$(*) \quad \Theta(SM) + \nu(SM, M) = \Theta(M)|SM,$$

where $\Theta(-)$ is the orientation cover of $(-)$, $\nu(SM, M)$ is the normal cover of SM in M and + denotes sum of isomorphism classes of Z_2-bundles.

In particular, if f is null-homotopic, we can take $SM = \emptyset$ and M becomes an oriented manifold.

Now Proposition 4.1 applies.

5. A FINAL EXTENSION AND EXAMPLES

Let $\mathfrak{a}$ be the category defined as follows. The objects of $\mathfrak{a}$ are linked resolutions; a map of resolutions $f : \rho \to \rho'$ $(\rho, \rho' \in \mathfrak{a})$ is a morphism of $\mathfrak{a}$ if $f = nf_0$, where n is a positive integer and f_0 is a quasi-linked map. Clearly f_0 and n are uniquely determined by f.

Let f be as above, then to each (ρ, n)-manifold M there is associated a (ρ', n)-manifold $f(M)$ defined by $f(M)$ = disjoint union of n copies of $f_0(M)$.

An $\mathfrak{a}$-system on a ball complex K is a covariant functor $S = S_K : \underset{\sim}{K} \to \mathfrak{a}$. Let $\mathfrak{M}_\sigma$ be the class of $S(\sigma)$-manifolds and $\mathfrak{M} = \mathfrak{M}(K) = \cup_\sigma \mathfrak{M}_\sigma$. If K is oriented then an (S, q)-mock bundle ξ^q over K consists of a function $\xi : K \to \mathfrak{M}$, such that, for each $\sigma \in K$, $\xi(\sigma)$ is an $(S(\sigma), q+\dim \sigma)$-manifold, labelled by σ, with

$$\partial\xi(\sigma) = \bigcup_{\tau < \sigma} [\sigma : \tau] S(\tau, \sigma)\xi(\tau)$$

where the obvious identifications are made in the union.

We note that an (S, q)-mock bundle does not have a total space or a projection. But nevertheless we can set up a theory of (S, q)-mock bundles in analogy to the usual case: The various notions of isomorphism, restriction, cobordism etc. for (S, q)-mock bundles are obtained by obvious modification from the definitions for ordinary mock bundles. We leave the reader to write down the details and to establish the existence of:

(i) an abelian group $\Omega^q(K, L; S)$, the q-th cobordism group of (K, L) with coefficients in S, which depends only on S and $|L| \subset |K|$.

(ii) a spectral sequence

$$H^p(|K|; \Omega_S^q) \Rightarrow \Omega^*(K; S).$$

The sheaf of local homology, considered in §4 for a $\mathbf{Z}_n$-manifold, provides examples of $\mathfrak{A}$ systems for many classes of polyhedra. Here we mention three:

(a) X = homology n-manifold, i.e. $H_i(X, X\text{-}x; \mathbf{Z}) = \tilde{H}_i(S^n; \mathbf{Z})$. Let $\rho_{\mathbf{Z}}$ be any short linked resolution of $\mathbf{Z}$. Then, for each $\sigma \in K$, $|K| = X$, we set $S(\sigma) = \rho_{\mathbf{Z}}$ and $S_{\rho_{\mathbf{Z}}}(\sigma < \tau) = m \circ \mathrm{id} : \rho_{\mathbf{Z}} \to \rho_{\mathbf{Z}}$ where $m : \mathbf{Z} \to \mathbf{Z}$ is the multiplication given by the local homology stack. Using the spectral sequence (ii) as in the proof of Theorem 2. 3, we are able to conclude that $\Omega^q(K; S_{\rho_{\mathbf{Z}}})$ is independent of the presentation $\rho_{\mathbf{Z}}$ and therefore provides a good definition of cobordism with coefficients in the sheaf of local n-homology of X.

(b) X = rational homology manifold. I.e. $H_i(X, X\text{-}x; \mathbf{Q}) = \tilde{H}_i(S^n; \mathbf{Q})$. In this case fix a linked resolution $\rho_{\mathbf{Q}}$ for $\mathbf{Q}$ and define $\Omega^q(K; S_{\rho_{\mathbf{Q}}})$ as in (a). Again $\Omega^q(K; S_{\rho_{\mathbf{Q}}})$ does not depend on the particular resolution.

(c) X is a polyhedron with only two intrinsic strata. Again there is a good definition of cobordism with coefficients in the sheaf of local homology. We leave details to the reader.

Finally, we conjecture that, at least when there is a 'good' definition of cobordism with coefficients in the sheaf of local homology, there is a 'Zeeman spectral sequence', cf. [1], relating bordism with cobordism with local coefficients. Proposition 4. 1 gives exactly this for a $\mathbf{Z}_n$-manifold.

REFERENCES FOR CHAPTER VI

[1] E. C. Zeeman. Dihomology III. Proc. Cam. Phil. Soc., (3) 13 (1963), 155-83.

VII · The geometry of CW complexes

In this final chapter we draw together all the ideas of the previous chapters by showing that an arbitrary cohomology theory is a geometric theory in an essentially unique way. Thus the geometric definitions of coefficients, operations etc. all apply to an arbitrary theory. This is achieved by examining the geometry of CW complexes. We will define a new concept, that of a transverse CW complex, which has all the geometric properties of ordinary cell complexes. In particular, it has a dual complex and transversality constructions can be applied. The transversality theorem (in §2) is a version for a CW complex of the theorem in part II §4. However the proof uses even less and is elementary!

If X is a based transverse CW complex and X^* its dual complex, then the subcomplex $\chi(X) \subset X^*$, consisting of dual objects other than the object dual to the basepoint, behaves with X exactly like the base of a Thom complex behaves with the whole complex. A map $f : M \to X$ can be made transverse to $\chi(X)$ (in fact the transversality theorem does exactly that) and the transverse map is determined by its values near $f^{-1}\chi(X)$. In this way, an arbitrary spectrum is seen to be a 'Thom' spectrum for a suitable theory (with singularities), see §§4 and 5.

Another consequence of this chapter is that CW complexes, already useful as homotopy objects, now have a beautiful intrinsic geometric structure. This has strong connections with stratified sets and the later work of Thom, see §3. We intend to write a paper [5] examining transverse CW complexes in greater detail and showing that they have all the properties enjoyed by cell complexes and semisimplicial complexes. In particular block bundles or mock bundles with base a transverse CW complex can be defined and have good geometric properties (see also §4 of this chapter).

The main theorem in §6 is that the stable homotopy category is equivalent to the category of geometric theories (theories of 'manifolds' with singularities, labellings etc., see Chapter IV) with 'resolutions' formally inverted. Thus any theory has cycles unique up to resolution of singularities, and any natural transformation of theories is equivalent to a relabelling followed by a 'resolution', see the examples in §6. Thus a theory has products if and only if the product of two cycles in the theory can be relabelled and then 'resolved' to give a cycle in the theory, see also the final remark of §6.

Throughout the chapter we will use the standard notation for CW complexes. Thus X, Y etc. denote CW complexes, e_1, e_2, e^i, e^j etc. denote cells; all cells have given characteristic maps denoted $h_1 : D_1 \to X$, $h_2 : D^j \to X$ etc., where $D^j = [-1, 1]^j \subset \mathbf{R}^j$. We denote $h_1(0) \in e_1 \subset X$ by $\hat{e}_1$ in analogy with the notation for the barycentre of a simplex. Other notation is in §1.

We would like to acknowledge a helpful conversation with C. T. C. Wall at the beginning of the work of this chapter.

1. BUILDINGS

In this section we describe a general structure which allows a dual structure to be defined. The examples will be used in later sections.

Let Y be a space and $\{b_i\}$ a partition of Y into disjoint subsets. Write $b_i < b_j$ if $i \neq j$ and $b_i \subset \bar{b}_j$. Define the simplicial complex sY to have for typical n-simplex a string

$$(b_{i_0} < b_{i_1} < \ldots < b_{i_n})$$

and faces given by omitting members of the string. Write $\breve{b}_i$ for the vertex (b_i) of sY and $\breve{b}_{i_0} \breve{b}_{i_1} \ldots \breve{b}_{i_n}$ for the simplex $(b_{i_0} < b_{i_1} < \ldots b_{i_n})$. sY has the structure of a cone complex in the sense of McCrory [3]; in particular we can define

$$C(\breve{b}_i) = cl\,\{\breve{b}_{i_0} \breve{b}_{i_1} \ldots \breve{b}_{i_n} \mid b_{i_n} = b_i\}$$

and

$$C^*(\breve{b}_i) = cl\,\{\breve{b}_{i_0}\breve{b}_{i_1} \ldots \breve{b}_{i_n} \mid b_{i_0} = b_i\}$$

in addition to the usual

$$st(\breve{b}_i,\ sY) = cl\,\{\breve{b}_{i_0}\breve{b}_{i_1} \ldots \breve{b}_{i_n} \mid b_i = b_{i_j} \ \text{ some } j\}$$

Then if we write

$$C(\breve{b}_i) = \breve{b}_i\dot{C}(\breve{b}_i), \quad C^*(\breve{b}_i) = \breve{b}_i\dot{C}^*(\breve{b}_i), \quad st(\breve{b}_i) = \breve{b}_i lk(\breve{b}_i),$$
$$\mathring{C}(\breve{b}_i) = C(\breve{b}_i) - \dot{C}(\breve{b}_i), \text{ etc.},$$

there is a canonical homeomorphism

$$\mathring{st}(\breve{b}_i) = \mathring{C}(\breve{b}_i) \times \mathring{C}^*(\breve{b}_i)\,.$$

A <u>plan</u> for Y is a map $d : Y \to sY$ such that $d^{-1}(\mathring{C}(\breve{b}_i)) = b_i$ and then $\{Y,\ \{b_i\},\ d\}$ is a <u>building</u> with <u>bricks</u> $\{b_i\}$ and plan d. We then have a new partition

$$dY \ \text{(the \underline{derived} of } Y) = \{d^{-1}(\mathring{\sigma}),\ \sigma \in sY\}$$

and the <u>dual building</u>

$$Y^* = \{Y,\ \{b_i^*\},\ d\} \ \text{ where } b_i^* = d^{-1}(\mathring{C}^*(\breve{b}_i)).$$

Notation. We write $\hat{b}_i$ for $d^{-1}(\breve{b}_i)$.

Examples 1.1. 1. Let K be a simplicial complex and $sK = dK$ the usual first derived of K. Then $\{K,\ \{\mathring{\sigma}\},\ \text{id.}\}$ is a building with bricks the open simplexes and plan the identity. Our notation is then consistent with the usual notation for barycentres and first deriveds.

2. Suppose Y is a CW complex in which the closure of each cell is a subcomplex. Let $\{e_i\}$ be the open cells of Y (which partition Y) then $\{Y,\ \{e_i\},\ d\}$ is a building, where d is defined by inductive conical extension. Then $\hat{e}_i$ is the centre of e_i and the notation is consistent.

3. Suppose $p : E \to K$ is a mock bundle, over a simplicial complex or a cone complex, in which each block has a collar and the projection is defined by mapping collar lines radially (the 'canonical projection'). Then $\{E, \{\text{open blocks}\}, p\}$ is a building where $sE \subset dK$ (the usual first derived). Note that if b_i is an open block of E then $\hat{b}_i = cl(b_i - \text{collar})$.

4. Suppose X is a stratified polyhedron in the sense of Stone [6]. I. e. X is partitioned by open manifolds, the strata, and provided with a system of regular neighbourhoods. There is also a local triviality condition which does not concern us. Then X defines a building with the strata for bricks in which the plan is obtained by using the mapping cylinder structure of a regular neighbourhood. Then $\hat{b}_i = cl(b_i - \text{neighbourhood of previous strata})$, the closed stratum.

2. TRANSVERSALITY

Transversality for CW complexes works nicely with either the smooth or pl categories. In this section we choose to work with the pl category but a similar treatment is possible with the smooth category.

Let M be a closed pl manifold and X a CW complex. A map $f : M \to X$ is transverse if for each cell $e_i \in X$ either $f^{-1}(e_i) = \emptyset$ or there is a commuting diagram

$$\begin{array}{ccc} cl(T_i) & \xrightarrow{\ t_i\ } & D_i \\ & {\scriptstyle f|}\searrow \quad \swarrow{\scriptstyle h_i} & \\ & X & \end{array}$$

where $T_i = f^{-1}(e_i)$, h_i is the characteristic map for e_i, t_i is the projection of a pl bundle (necessarily trivial) and $cl(T_i)$ has codimension zero in M. Notice that this implies that $\hat{T}_i = t_i^{-1}(0)$ is a submanifold of M of codimension i. Notice also that if $X = X_0 \cup e_i$ then $M = M_0 \cup \hat{T}_i \times D_i$, where $\hat{T}_i \times D_i$ has the form of a 'generalised handle' attached to M_0 by $\hat{T}_i \times \partial D_i$.

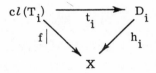

In the case that $\partial M \neq \emptyset$, we insist that $cl(T_i)$ meets ∂M in a subtube $\hat{T}' \times D_i$ where $\hat{T}' \subset \partial\hat{T}_i$ has codimension 0. Thus $f|\partial M$ is also transverse.

We say that the CW complex X is *transverse*, or that X is a *TCW complex*, if each attaching map is transverse to the skeleton to which it is mapped.

Transversality theorem 2.1. *Suppose X is a TCW complex and* $f : M \to X$ *a map, where M is a compact pl manifold. Suppose* $f|\partial M$ *is transverse. Then there is a homotopy of f rel* ∂M *to a transverse map.*

Corollary 2.2. *Any CW complex gives rise to a TCW complex of the same homotopy type obtained by inductively homotoping attaching maps to make them transverse.*

Proof of the transversality theorem. Since im(f) is contained in a finite subcomplex of X, we can assume X is finite and proceed by induction on the number of cells of X. Suppose $X = X_0 \cup e_i$.

Choose a collar c for ∂M in M and by a preliminary homotopy rel ∂M assume that f is constant on collar lines. Now apply the standard transversality theorem (see Remark 2.3 below) to make f transverse to $\hat{e}_i$ in e_i, by a homotopy of f rel im(e). We now have a diagram

$$\begin{array}{ccc} T & \xrightarrow{\ t\ } & D_i(\varepsilon) \\ & {\scriptstyle f|}\searrow \quad \swarrow{\scriptstyle h_i|} & \\ & X & \end{array}$$

where $D_i(\varepsilon)$ is a small disc in D_i centred on 0 and t is a trivial bundle. Moreover $f|\text{im}(c)$ is already transverse to the whole of e_i. Hence, by composing $f|cl(M - \text{im}(c))$ with a standard homotopy of $\bar{e}_i$ in itself (obtained by expanding $D_i(\varepsilon)$ onto D_i and using h_i) and then extending, by using the collar in the usual way, to a homotopy of f which keeps ∂M fixed, we have the same diagram with D_i replacing $D_i(\varepsilon)$.

Now write $\delta T = \hat{T} \times \partial D_i$ ($\hat{T} = t^{-1}(0)$ as usual), $M_0 = cl(M - T)$, $M_1 = cl(\partial M - T)$. Then M_0 is a compact manifold with boundary $M_1 \cup \delta T$. Now from definitions and the diagram both $f|M_1$ and $f|\delta T$

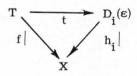

are transverse to X_0 (and to X). Thus $f | \partial M_0$ is transverse to X_0 and by induction we may homotope f further rel ∂M_0 to make it transverse, as required.

Remark 2.3. We only used the simplest pl transversality theorem, which has an elementary proof using the fact that the preimage of the barycentre of a top dimensional simplex by a simplicial map is framed - a fact that has been known for about forty years! The corresponding smooth theorem is also elementary, using Sard's theorem.

Now let $M_+ = M \underset{\partial M \times 0}{\cup} \partial M \times I$, i.e. M with a collar glued on 'on the outside', and let $q : M_+ \to M$ be the map which projects the collar back onto ∂M. If $f : M \to X$ is a map, then define $f_+ : M_+ \to X$ to be $f \circ q$. We say $f : M \to X$ is <u>weakly transverse</u> if f_+ is transverse. Thus $f | \partial M$ is transverse but some of the tubes T_i might only have codimension 0 in ∂M. An example of a weakly transverse map is a characteristic map for a cell in a TCW complex. In fact it can easily be seen that X is a TCW complex if and only if all the characteristic maps are weakly transverse.

A map $f : X \to Y$ between TCW complexes is <u>transverse</u> if $f \circ h_i : D_i \to Y$ is weakly transverse for each characteristic map h_i of X.

Theorem 2.4. <u>There is a category TCW consisting of TCW complexes and transverse maps. The inclusion TCW ⊂ CW induces an equivalence of homotopy categories. TCW is closed under cross product, wedge product, factorisation of a subcomplex and smash product. In particular it is closed under suspension (smash product with S^1).</u>

Proof. That TCW is a category follows from 2.5 below and definitions. That TCW ⊂ CW induces an equivalence of homotopy categories follows from 2.2 and 2.6 below. The rest is a matter of trivial verification.

Lemma 2.5. $f : M \to X$, $g : X \to Y$ <u>are both transverse maps where</u> M <u>is a compact</u> pl <u>manifold and</u> X, Y <u>are</u> TCW <u>complexes. Then</u> $g \circ f : M \to Y$ <u>is transverse.</u>

Corollary 2.6. Any map between TCW complexes is homotopic to a transverse map. The homotopy can be chosen to keep fixed a subcomplex on which the map is already transverse.

Proof. Use the transversality theorem inductively to shift cells to make the map transverse. The lemma and induction ensure that a cell is already transverse on its boundary.

The proof of Lemma 2.5 uses the description of M as a framified set which is given in the next section and will thus be left for convenience until the end of that section.

3. FRAMIFIED SETS

Roughly speaking, a framified set is a stratified set in the sense of Stone [6] in which all the block bundles are trivialised. The precise definition is similar to the definition of killing in Chapter IV, and in fact a precise connection will be formulated in §4. The definition is by induction on the length of filtration.

Definition 3.1. A framified set $\underset{\sim}{X}$ of length n consists of

(1) A polyhedron X with a filtration

$$\{X = X_1 \supset X_2 \supset \ldots X_n \supset X_{n+1} = \emptyset\}$$

such that $X_i - X_{i+1}$ is a manifold for $i = 1, 2, \ldots, n$.

(2) A regular neighbourhood system for the filtration, $N_{i,j}$, $1 \leq j \leq i \leq n$, (see Remark 3.2 below).

(3) For each $i \leq n$ a framified set $\underset{\sim}{L}_i$ of length $i - 1$.

(4) For each $i \leq n$ an isomorphism of filtered sets

$$h_i : \{N_{i,1} \supset N_{i,2} \supset \ldots N_{i,i}\} \cong N_{i,i} \times \underset{\sim}{C}_i$$

where $\underset{\sim}{C}_i = \{C(L_{i,1}) \supset \ldots C(L_{i,i-1}) \supset \text{pt.}\}$ is the cone on $\underset{\sim}{L}_i$ with the cone point added as the final stage of the filtration.

The cone flag $\underset{\sim}{C}_i$ is a fibre or model for the framified set $\underset{\sim}{X}$, and the framified set $\underset{\sim}{L}_i$ is a link for $\underset{\sim}{X}$. There are obvious notions of restriction of a framified set to a suitable subpolyhedron and of product of a framified set with a manifold. A pl homeomorphism is an isomorphism of framified sets if it commutes with all the extra structure. In particular two isomorphic framified sets have the same (or identifiable) system of links. The final condition is:

(5) h_i restricts to an isomorphism of framified sets

$$\{\dot{N}_{i,1} \supset \dot{N}_{i,2} \supset \ldots \dot{N}_{i,i-1}\} \cong N_{i,i} \times \underset{\sim}{L}_i \ .$$

Remark 3.2. A regular neighbourhood system is constructed inductively by defining $N_{n,n} = X_n$ and $N_{n,j}$ is a simultaneous system of second derived neighbourhoods of X_n in X_j. Then define $X_i' = X - \text{int}(N_{n,i})$ and proceed with the construction for

$$X_1' \supset X_2' \supset \ldots \supset X_{n-1}' \supset \emptyset.$$

Existence and uniqueness of regular neighbourhood systems thus follows from the usual regular neighbourhood theorem.

Notice that a framified set is a building with bricks the strata $\{X_i - X_{i-1}\}$ and plan defined by using the cone structures (see §1 Example 4). We are not interested in the specific ordering of the strata of $\underset{\sim}{X}$ but only in the partial ordering given by the geometric structure of X, and we will allow an isomorphism of framified sets to change the order. The importance of framified sets lies in the following theorem, which is essentially an observation.

Theorem 3.3. Let $f : M \to X$ be a transverse map to a TCW, then M has the structure of a framified set determined up to isomorphism by the map f.

Proof. Since the image of f is contained in a finite subcomplex of X we can assume without loss that X is finite and proceed by induction on the number of cells of X. In fact we will produce a framification of the same length as the number of cells. Let $X = X_0 \cup e$. Define $M_t = f^{-1}(\hat{e})$ and $N_{t,1} = cl(f^{-1}(e))$ (i.e. $M_t = \hat{T}$ and $N_{t,1} = T$ in the notation of the previous section). Now let $M_0 = cl(M - T)$ then M_0 and $\delta T \cong \hat{T} \times S_{j-1}$ have the structure of framified sets by induction.

Moreover, by uniqueness we can choose the structure on δT to be the product of the framification of S_{j-1} with $\hat{T}$. Choose the indexing to agree and extend the strata to M by adjoining the 'cones' on their intersection with δT. I. e.

$$M_i = (M_0)_i \cup \hat{T} \times (C(S_{j-1})_i - \text{cone pt.}).$$

Finally the isomorphism h_t is provided by the chosen product structure on T. Induction now gives all the structure of a framified set to M. Uniqueness is clear.

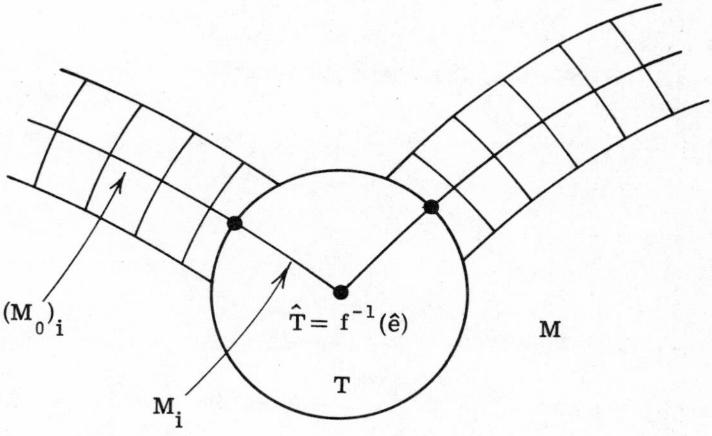

Fig. 20

We can also describe this framification (at least as a building) quickly as follows. Take the dual complex to X and let

$$X = X_1^* \supset X_2^* \supset \dots \supset X_t^* = \hat{e}$$

be the corresponding filtration of X. Then $M_t = f^{-1}(X_t^*)$. In other words the building is the pull-back by f of the dual building to X.

Notice that the 'models' $\underset{\sim}{C}_i$ are just the closures of the cone flags

$$e_i \cap (X_1^* \supset X_2^* \supset \dots X_i^*) .$$

Since the models depend only on X, we say that the framification of M is modelled on X. It is clear that there is a 1 - 1 correspondence between transverse maps $f : M \to X$ and framifications of M modelled on X. This observation will be developed in the next section when we 'classify' homotopy classes of maps from one TCW to another. To end this section we will prove the lemma left at the end of the last section.

First observe that, by inductively choosing collars on the frontiers of the $N_{i,j}$ we can find an isomorphic system with smaller closed strata and larger cone flags (cf. Stone on 'minidivision' [6]); this has the effect, for the framification given by a transverse map $f : M \to X$, of replacing the neighbourhoods $\hat{T}_i \times D_i$ by neighbourhoods of the form $\hat{T}_i' \times D_i^+$ where $\hat{T}_i' \cong \hat{T}_i$ and $D_i^+ = D_i \cup_\partial$ collar, and it is not hard to see that we can choose all the collars so that there are diagrams

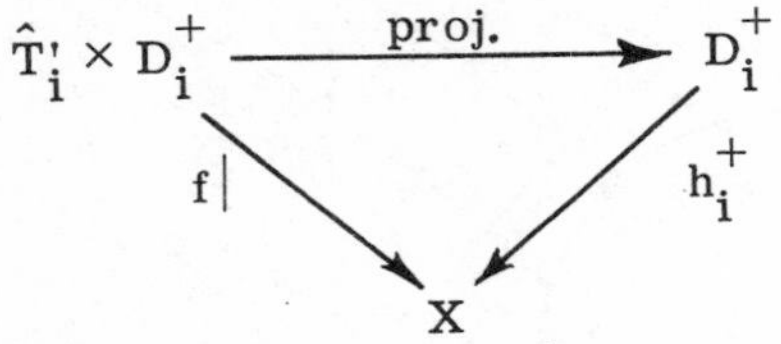

We call this framification an extension of the original one.

Proof of Lemma 2.5. Let e_k be a cell in Y and $T_k = (g \circ f)^{-1} e_k$. We have to show that $cl(T_k)$ is a product $\hat{T}_k \times D_k$ of codimension zero in M which meets ∂M in a similar subproduct. Choose an extended framification for $f : M \to X$. This means that we can regard M as made of generalised handles of the form $\hat{T}_i' \times D_i^+$. Since $g \circ f | : D_i^+ \to Y$ is transverse (from the definition of transversality for TCW's) we have $D_i^+ \cap T_k = Q \times D_k$ say where Q is a manifold with boundary and $Q \times D_k$ meets ∂D_i^+ in a similar subtube. Here we are regarding D_i^+ as included in M as pt. $\times D_i^+ \subset \hat{T}_i' \times D_i^+$. Thus

$$(\hat{T}_i' \times D_i^+) \cap T_i = (Q \times \hat{T}_i') \times D_k$$
$$= Q' \times D_k \text{ say.}$$

But all the product structures are coherent in the D_k factor and the required product structure on T_i is seen.

4. MANIFOLDS AND MOCK BUNDLES MODELLED ON X

We now rephrase the results of §3 by omitting the first stratum throughout. The idea is to obtain a formulation which is invariant under suspension and thus carries over to stable maps.

Let X be a based TCW. Let $* \in X$ be the basepoint (regarded as the first cell in X) and let $\chi(X)$ be the subcomplex of X^* consisting of the duals to cells other than $*$. If $f : M \to X$ is transverse then the associated stratification of M starts $M \supset \chi(M, f)$ where $\chi(M, f) = f^{-1}(\chi(X))$. Let e_i be a typical cell of X. Collapse the notation of §3 and write $C_i = cl(\chi(X) \cap e_i)$, then C_i is the cone $\hat{e}_i L_i$, where L_i is the link associated to e_i, and we label the cone point by e_i. L_i is in fact a framified set embedded in $S_i = \partial D_i$. The set of basic links $\{L_i\}$ defines a theory of manifolds with singularities (see IV §3) called _free X manifolds._ Using the fact that $L_i \subset S_i$, we have an intuitive notion of a _framed_ X manifold. The precise formulation is in terms of killing as in IV §4. Suppose X is finite and $X = X_0 \cup e_i$. Suppose inductively that framed X_0 manifolds have been defined so that L_i is a framed X_0 manifold. The theory of framed X manifolds is the theory obtained from this theory by killing L_i and labelling the new stratum of singularities by e_i. In general define a framed X manifold to be a framed X' manifold where $X' \subset X$ is a finite subcomplex. From the definitions of killing and framified sets, it is easy to see that a framification of M modelled on X is equivalent to a framed X manifold embedded in M. From Theorem 3.3 and the transversality theorem we have:

Proposition 4.1. _There is a 1 - 1 correspondence between the set of homotopy classes of maps_ $[M, X]$ _and the set of cobordism classes of framed_ X _manifolds embedded in_ M. _In particular_ $\pi_n(X) \cong \Omega^n_X$ _where_ Ω^n_X _means the group of cobordism classes of framed_ X _manifolds embedded in_ S^n.

We can extend the proposition to maps $[Y, X]$ where Y is an unbased TCW using an extension of mock bundles to TCW's. Let Y be a TCW then a subset $E \subset Y$ is the total space of an *embedded mock bundle* (of dimension $-q$) provided that for each cell $e_i \in Y$ there is a diagram

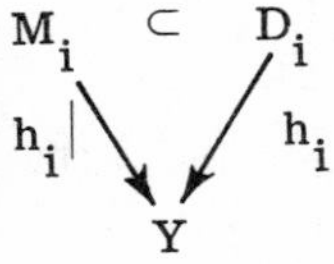

where $M_i = h_i^{-1}(E)$ and is a proper submanifold of D_i of codimension q. The following proposition is a generalisation of Lemma 1.2 of Chapter II and is proved by a similar argument to Lemma 2.5. We omit the details.

Proposition 4.2. *Let* $E \subset Y$ *be an embedded mock bundle and* $f : M \to Y$ *a transverse map, then* $f^{-1}(E)$ *is a proper submanifold of* M *of codimension* q.

The proposition implies that mock bundles can be pulled back and hence give a contravariant functor on TCW.

There is an obvious extension of the notion of mock bundle to mock bundle with singularities and framed mock bundle. The next proposition is, like 4.1, essentially an observation:

Proposition 4.3. *There is a* $1 - 1$ *correspondence between transverse maps* $Y \to X$ *and framed* X *mock bundles embedded in* Y. *Hence* $[Y, X] \cong \Omega_X(Y)$, *where* $\Omega_X(Y)$ *denotes cobordism classes of framed* X *mock bundles in* Y.

There is a based version of 4.3; $[Y, X]_* = \tilde{\Omega}_X(Y)$ where $\tilde{\Omega}_X(Y)$ denotes cobordism classes of framed X mock bundles in $Y - *$.

We will now stabilise these results. Let SX denote the (reduced) suspension of X which is again a TCW then $\chi(SX) = \chi(X)$ and if $f : S^n \to X$ is transverse then so is $Sf : S^{n+1} \to SX$. Moreover $\chi(S^{n+1}, Sf) = \chi(S^n, f)$ and the only difference between the framification of S^{n+1} given by Sf and that of S^n by f is that all the bundles are enlarged by adding a trivial 1-disc bundle. We thus have the stable version of 4.1 and 4.3 (for details of the stable category see Adams [1]):

Proposition 4.4. There is a 1 - 1 correspondence between stable homotopy classes of (based) maps $S^n \to X$ and cobordism classes of stably framed X manifolds. There is a 1 - 1 correspondence between stable homotopy classes of stable maps $\{Y, X\}$ and framed X mock bundles over $Y - *$ (i.e. embedded in $S^\infty Y - *$).

Remarks. 1. The above notion of a mock bundle over Y is a direct generalisation of the definition for cell complexes in Chapter II.

2. We have been deliberately careless (or rather uninformative) about dimensions; this will be remedied in the next section.

3. See Example 2 at the end of the next section for a clarification of the relation between this representation and 'killing'.

5. THE CYCLES OF A HOMOLOGY THEORY

At the end of the last section we had represented a cohomology theory, which came from the suspension of a CW complex, as a mock bundle theory. In this section we extend this result to arbitrary spectra and observe that the corresponding homology theory is the corresponding bordism theory. The results are most elegant for connected spectra when we will be able to represent $h_n(\ )$ by n-manifolds with singularities. For non-connected spectra we will need manifolds of non-constant dimension. Uniqueness of representatives will be discussed in §6.

We follow Adams' treatment of the stable category, [1].

A spectrum $\underset{\sim}{X}$ is a sequence $\{X_i, q_i, i \geq 0\}$ of based CW complexes and based maps where $q_i : SX_i \to X_{i+1}$ is an isomorphism onto a subcomplex. It is connected if X_i has no cells other than $*$ in dimensions $< i$.

Suppose $\underset{\sim}{X}$ is connected and, for some $i > 0$, $X_i = * \cup e_1^i \cup e_2 \cup \ldots$ Make X_i into a TCW and, by further homotopies, ensure that each cell $e_2, e_3 \ldots$ wraps non-trivially around e_1^i (geometrically, not homotopically!). Then $\chi(X_i) = \hat{e}_1 \cup \ldots$ will have constant codimension i in X_i. Now SX_i is again transverse and since $SX_i \subset X_{i+1}$ we can make X_{i+1} transverse keeping SX_i fixed and, by further homotopies, ensure that each cell in $X_{i+1} - SX_i$ wraps non-trivially around Se_1. Thus $\chi(X_{i+1})$ again has constant codimension (this time $i + 1$). Proceeding in this way we can make each X_j transverse, for $j \geq i$, with $SX_j \subset X_{j+1}$ and $\chi(X_j)$ of constant codimension j.

Now define the geometric homology theory associated to $\underset{\sim}{X}$ by taking for basic links all the links defined by X_j for $j \geq i$, with the obvious identifications given by the inclusions $SX_j \subset X_{j+1}$. Each link is stably framed and it makes sense to talk of stably framed $\underset{\sim}{X}$ manifolds and from the results of §4, Chapters II and IV and definitions we have:

Theorem 5.1. The homology and cohomology theories defined by the spectrum $\underset{\sim}{X}$ are equivalent to the bordism and mock bundle theories based on stably framed $\underset{\sim}{X}$ manifolds. Moreover by choosing $\chi(X_j)$ to have constant codimension j we have ensured that $h_n(\ ;\underset{\sim}{X})$ is represented by $\underset{\sim}{X}$ manifolds of (genuine) dimension n and that $h^q(\ ;\underset{\sim}{X})$ is represented by mock bundles of fibre dimension q.

Remark 5.2. The first half of the theorem is true for non-connected spectra by the same proof but since $\chi(X_j)$ will in general have varying codimension in X_j, the cycles will be of mixed dimension, possibly of unbounded dimension. See the examples below.

Examples 5.3. 1. $X_j = S^j$. $\chi(X_j) = \text{pt.}$ and an $\underset{\sim}{X}$ manifold is just a framed manifold. Thus we have the usual representation for stable (co)homotopy.

2. For some j, $X_j = S^j \cup_f D^k$ where $f : S^{k-1} \to S^j$ is a given map and

$$X_i = S^i,\ i < j,$$
$$X_{l+j} = S^l X_j .$$

Then

$$L_1 = \emptyset,\ C_1 = \text{pt. (of codim j)},$$
$$L_2 = M^{k-j-1},\ C_2 = C(M)$$

where M is the framed submanifold corresponding to f. $\underset{\sim}{X}$ manifolds have two strata both framed and the neighbourhood of the smaller stratum in the larger stratum is a product with $C(M)$.

In other words $\underset{\sim}{X}$ theory is framed bordism with M 'killed', see Chapter IV §4. This example is the germ of the whole construction. Each new cell attached determines a manifold in the previous theory, which, when killed, gives the new theory.

3. The case $f \simeq *$ of the example 2 is also worth discussing in detail. If $f = *$, i.e. $X_j = S^j \vee S^k$, then the theory is a 'mixed dimension' theory, i.e. an 'n-cycle' is the union of a framed n-manifold with a framed $(n+j-k)$-manifold. If $f \neq *$ then M is a framed manifold bordant to $\emptyset$. Thus X theory is not a mixed dimension theory but it has a resolvable singularity: there is an elementary resolution connecting the two theories, see Example 6.4(4).

4. X = Thom spectrum $\underset{\sim}{MPL}$ or $\underset{\sim}{MO}$ etc. MPL_n has a cell structure in which each cell is of the form (cell of BPL_n) $\times \mathring{I}^n$. Then if BPL_n is a TCW, so is MPL_n and $\chi(MPL)_n = BPL_n$. Thus $\chi(M, f)$ is just a submanifold of M of codimension n and we recover the usual representation of bordism. Similarly with $\underset{\sim}{MO}$ we get submanifolds with normal vector bundles, i.e. smooth(able) submanifolds. In this example the singularities are 'virtual', they are of the form C(framed sphere) and serve to allow the normal bundle of $\chi(M, f)$ to be non-trivial.

5. $\underset{\sim}{X}$ = Moore spectrum. E.g. for Z_n, $X_j = S^j \cup_n D^{j+1}$. Then $L_1 = \emptyset$, L_2 = n points and we get framed Z_n manifolds. In general the geometric description fits with that given in Chapters III and IV (for stable homotopy, but see also Example 8 below).

6. If $\underset{\sim}{X}$ and $\underset{\sim}{Y}$ are TCW spectra then so is $\underset{\sim}{X} \vee \underset{\sim}{Y}$ and an $\underset{\sim}{X} \vee \underset{\sim}{Y}$ manifold is merely the union of an $\underset{\sim}{X}$ manifold with a $\underset{\sim}{Y}$ manifold. Take care about dimensions, see Example 3 above.

7. If $\underset{\sim}{X}$ and $\underset{\sim}{Y}$ are TCW spectra then so is a naive smash product $\underset{\sim}{X} \wedge \underset{\sim}{Y}$, see [1; p. 161], and a cell of $\underset{\sim}{X} \wedge \underset{\sim}{Y}$ is the product of one of $\underset{\sim}{X}$ with one of $\underset{\sim}{Y}$ and has for link the join of the corresponding links

for $\underset{\sim}{X}$ and $\underset{\sim}{Y}$. Thus an $\underset{\sim}{X} \wedge \underset{\sim}{Y}$ manifold is a manifold with singularities being joins of $\underset{\sim}{X}$ singularities and $\underset{\sim}{Y}$ singularities.

8. Combine Examples 4, 5 and 7. Then bordism with coefficients has exactly the description given in part III!

6. RESOLUTION OF SINGULARITIES AND THE MAIN THEOREM

Let T be a geometric homology theory and suppose that in T there are two singularities of the form

(1) C(M)

(2) $C(C(M) \cup_M W)$ where W is a bordism (in T) of M to $\emptyset$.

Suppose also that C(M) does not appear in any other basic link. Then a T-manifold can be resolved so as to delete both of these two singularities. This is done by resolving all the C(M) type singularity using (2) which is essentially a bordism in T of C(M) to a T-manifold with no C(M) singularity. The method is similar to the proof of exactness in IV Proposition 4.1. Once there are no C(M) singularities, then there can be none of the second type either. The precise description of this resolution process is contained in the CW interpretation which follows. Let T' be the theory in which these two singularities do not appear then we say that T' is an <u>elementary resolution</u> of T. A simultaneous family of elementary resolutions will also be called an elementary resolution. A <u>resolution</u> is a countable sequence of elementary resolutions.

Now suppose that X and X' are TCW's and T and T' above are the corresponding theories. Then $X' \subset X$ and X differs from X' by the addition of two cells e^n and e^{n+1} with e^{n+1} attached by degree 1 on e^n and e^n otherwise free. In other words X' differs from X by an <u>elementary collapse.</u> Thus the geometric analogue of collapsing is resolution and the process of resolution of a given T-manifold described at the beginning of the section is just transversely deforming the map $S^n \to X$ (which defines the manifold) into X', using the deformation retraction $X \to X'$. The inclusion $X' \subset X$ is an <u>elementary expansion.</u> An <u>expansion</u> is a countable sequence of (simultaneous families of) elementary expansions.

The following theorem is proved by an argument similar to that contained in Chapter I (in fact rather simpler). * Its proof will therefore be omitted.

Theorem 6.1. Let $\underline{\underline{CW}}$ and $\underline{\underline{SCW}}$ denote the categories of CW complexes and isomorphisms onto subcomplexes and CW spectra and isomorphisms onto subspectra respectively. Let Σ denote the expansion in $\underline{\underline{CW}}$ or $\underline{\underline{SCW}}$. Then there are isomorphisms

$$h\underline{\underline{CW}} \cong \underline{\underline{CW}}(\Sigma^{-1})$$
$$h\underline{\underline{SCW}} \cong \underline{\underline{SCW}}(\Sigma^{-1}).$$

Combining with 2.4 we have:

Corollary 6.2. There are equivalences of categories

$$h\underline{\underline{CW}} \simeq \underline{\underline{TCW}}(\Sigma^{-1})$$
$$h\underline{\underline{SCW}} \simeq \underline{\underline{STCW}}(\Sigma^{-1}).$$

Now let $\mathcal{G}$ denote the category whose objects are geometric theories and whose morphisms are inclusions of theories or 'relabellings'. We now regard all theories as theories of framed manifolds with singularities, non-framed manifolds being dealt with as in Example 5.3(4), by allowing singularities corresponding to the twisting of the normal bundle. A theory T is included in a theory T' if the links of T' include those of T up to a relabelling. Let R denote the resolutions in $\mathcal{G}$. From the discussion at the beginning of this section and 5.1 we have an isomorphism

$$\mathcal{G}(R) \cong \underline{\underline{STCW}}(\Sigma^{-1}).$$

Combining this with 6.2 we have:

Main theorem 6.3. There is an equivalence of categories

$$\mathcal{G}(R) \simeq h\underline{\underline{SCW}}.$$

---

* There are similar set theoretic problems to those encountered in Chapter I §4. These can be dealt with in a similar way.

In other words the stable homotopy category is equivalent to the category of geometric theories with resolutions formally inverted. The operations $\vee$ and $\wedge$ in $h\underline{\underline{SCW}}$ are described geometrically as 'union' and 'join' of singularities, see 5. 3(6) and (7).

Remarks and corollaries 6. 4. 1. There is an unstable version, using 'embedded' geometric theories (corresponding to the embedded mock theories of §4). We obtain an equivalence $E\mathcal{G}(R) \simeq h\underline{\underline{CW}}$.

2. Theorem 6. 3 gives the answer to the problem of uniqueness of geometric representatives for a given homology theory. Two geometric theories give equivalent homology theories if and only if they differ by a sequence of resolutions and their inverses.

3. The theorem also describes the stable maps geometrically (i. e. natural transformations of theories, operations etc.). Such a map always has the form inclusion (relabelling) followed by resolution. This follows from an analogue of Lemma 2. 1 of Chapter I. See also the next two examples.

4. The example described in 5. 3(3) fits into the setting of this section as follows. Let $X' = S^j \cup_H (D^k \times I)$ where H is the homotopy of f to $*$. Make X' transverse relative to the two 'ends'. Then X' defines a geometric theory of which both the theories described in 5. 3(3) are elementary resolutions. The new singularity in X' is $C(C(M) \cup_\partial W \cup \text{pt.})$ and can be used to resolve either the lower dimensional piece into a singularity of the higher dimensional piece, or conversely. This example makes it clear how resolutions can change the appearance (and dimension) of a manifold drastically.

5. If we now consider the map

$$S^j \cup_f D^k \to S^k$$

given by collapsing S^j to a point, then the geometric description of this map is 'restrict to the singularity'. This is seen by including $S^j \cup_f D^k$ in $D^{j+1} \cup_f D^k$ which collapses to S^k and using an argument like Example 4. This example makes it clear that 'restriction to a singularity' is an operation and in fact many classical operations have this form, (cf. McCrory [4]).

6. Finally a theory with products is one in which there is a map $\underset{\sim}{X} \wedge \underset{\sim}{X} \to \underset{\sim}{X}$ with suitable properties. By expanding $\underset{\sim}{X}$ by a sequence of mapping cylinders we can replace this by an equivalent inclusion $\underset{\sim}{X}' \wedge \underset{\sim}{X}' \subset \underset{\sim}{X}'$. Now if $\underset{\sim}{S}^n \to \underset{\sim}{X}'$, $\underset{\sim}{S}^m \to \underset{\sim}{X}'$ represents two manifolds in the theory then $\underset{\sim}{S}^n \wedge \underset{\sim}{S}^m \to \underset{\sim}{X}' \wedge \underset{\sim}{X}'$ represents their geometric product (see 5.3 Example 7). Thus a theory has products if and only if it is possible to find a geometric representation in which the product of two cycles is again a cycle (after possibly relabelling). Thus the classical examples of ring theories (bordism etc.) are essentially the general example! Note, however, the case of R-bordism, see Chapter III, in which the ring structure appears naturally via a resolution of singularities.

REFERENCES FOR CHAPTER VII

[1] J. F. Adams. Stable homotopy and generalised homology. Chicago U. P. (1974).

[2] P. Gabriel and M. Zisman. Homotopy theory and the calculus of fractions. Springer-Verlag, Berlin (1967).

[3] C. McCrory. Cone complexes and duality, (to appear).

[4] C. McCrory. Geometric homology operations, (to appear).

[5] C. P. Rourke and B. J. Sanderson. CW complexes as geometric objects, (to appear).

[6] D. A. Stone. Stratified polyhedra. Springer-Verlag lecture notes No. 252.